Introduction to Crystallography

Introduction to Crystallography

Marina Wright

Willford Press
www.willfordpress.com

Published by Willford Press,
118-35 Queens Blvd., Suite 400,
Forest Hills, NY 11375, USA

ISBN: 978-1-64728-002-4

Cataloging-in-Publication Data

Introduction to crystallography / Marina Wright.
p. cm.
Includes bibliographical references and index.
ISBN 978-1-64728-002-4
1. Crystallography. 2. Mineralogy. 3. Chemistry. I. Wright, Marina.
QD905.2 .I58 2022
548--dc23

For information on all Willford Press publications
visit our website at www.willfordpress.com

Willford Press

Contents

Preface

Crystal, also known as crystalline solid, is a material whose constituents are arranged in a highly ordered microscopic structure. It forms a crystal lattice that extends in all directions. The branch of science, which studies crystals, its formation, and the bonding and arrangement of atoms in crystalline solids is known as crystallography. Crystal structure is the ordered arrangement of atoms, molecules and ions in a crystal. It is used in materials science to characterize different materials and phase identification. X-ray crystallography is a method that determines the molecular conformations of large biomolecules, particularly nucleic acids such as RNA and DNA, and protein. Neutron crystallography is used in refining structures obtained by X-ray methods. Electron crystallography uses transmission electron microscope to determine the arrangement of atoms in solids. This book provides comprehensive insights into the field of crystallography. Most of the topics introduced herein cover new techniques and the applications of crystallography. This book will serve as a reference to a broad spectrum of readers.

To facilitate a deeper understanding of the contents of this book a short introduction of every chapter is written below:

Chapter 1- A crystal is a type of solid whose molecules, atoms and ions are organized in a highly ordered microscopic structure that forms a crystal lattice. Crystal habit deals with the specific orientations and geometrical shapes that a crystal exhibits. This is an introductory chapter which will briefly introduce all the significant aspects of crystal.

Chapter 2- Crystallography is the branch of chemistry which is concerned with the arrangement of atoms in a crystal. Symmetry in crystallography is useful for the characterization of different crystals and its physical properties such as thermal conductivity and optical activity. This chapter has been carefully written to provide an easy understanding of the varied facets of crystallography.

Chapter 3- There are various types of crystallography such as quantum crystallography, x-ray crystallography, electron crystallography, optical crystallography, isometric crystallography and tetragonal crystallography. Fourier space, Bragg's law of diffraction, etc. are some of the aspects that fall under its domain. All these different types of crystallography and related aspects have been carefully analyzed in this chapter.

Chapter 4- A crystal structure is defined as the ordered arrangements of atoms, ions and molecules in a crystal. The space groups, lattices, point groups, crystal forms, etc. together form a crystal system. This chapter closely examines these concepts of crystal structure and crystal system to provide an extensive understanding of the subject.

Chapter 5- In crystallography, crystal defects are the discrepancies in the regular pattern, arrangement and organization of atoms of crystalline solids. Some of the defects are intrinsic and extrinsic defects, Frenkel defects, interstitial defects, line defects, surface defects, volume defects, etc. The topics elaborated in this chapter will help in gaining a better perspective about these crystal defects.

I owe the completion of this book to the never-ending support of my family, who supported me throughout the project.

Marina Wright

1

Introduction to Crystal

A crystal is a type of solid whose molecules, atoms and ions are organized in a highly ordered microscopic structure that forms a crystal lattice. Crystal habit deals with the specific orientations and geometrical shapes that a crystal exhibits. This is an introductory chapter which will briefly introduce all the significant aspects of crystal.

Crystal is any solid material in which the component atoms are arranged in a definite pattern and whose surface regularity reflects its internal symmetry.

Classification

The definition of a solid appears obvious; a solid is generally thought of as being hard and firm. Upon inspection, however, the definition becomes less straightforward. A cube of butter, for example, is hard after being stored in a refrigerator and is clearly a solid. After remaining on the kitchen counter for a day, the same cube becomes quite soft, and it is unclear if the butter should still be considered a solid. Many crystals behave like butter in that they are hard at low temperatures but soft at higher temperatures. They are called solids at all temperatures below their melting point. A possible definition of a solid is an object that retains its shape if left undisturbed. The pertinent issue is how long the object keeps its shape. A highly viscous fluid retains its shape for an hour but not a year. A solid must keep its shape longer than that.

Basic Units of Solids

The basic units of solids are either atoms or atoms that have combined into molecules. The electrons of an atom move in orbits that form a shell structure around the nucleus. The shells are filled in a systematic order, with each shell accommodating only a small number of electrons. Different atoms have different numbers of electrons, which are distributed in a characteristic electronic structure of filled and partially filled shells. The arrangement of an atom's electrons determines its chemical properties. The properties of solids are usually predictable from the properties of their constituent atoms and molecules, and the different shell structures of atoms are therefore responsible for the diversity of solids.

All occupied shells of the argon (Ar) atom, for example, are filled, resulting in a spherical atomic shape. In solid argon the atoms are arranged according to the closest packing of these spheres. The iron (Fe) atom, in contrast, has one electron shell that is only partially filled, giving the atom a net magnetic moment. Thus, crystalline iron is a magnet. The covalent bond between two carbon (C) atoms is the strongest bond found in nature. This strong bond is responsible for making diamond the hardest solid.

Long- and Short-range Order

A solid is crystalline if it has long-range order. Once the positions of an atom and its neighbours are known at one point, the place of each atom is known precisely throughout the crystal. Most liquids lack long-range order, although many have short-range order. Short range is defined as the first- or second-nearest neighbours of an atom. In many liquids the first-neighbour atoms are arranged in the same structure as in the corresponding solid phase. At distances that are many atoms away, however, the positions of the atoms become uncorrelated. These fluids, such as water, have short-range order but lack long-range order. Certain liquids may have short-range order in one direction and long-range order in another direction; these special substances are called liquid crystals. Solid crystals have both short-range order and long-range order.

Solids that have short-range order but lack long-range order are called amorphous. Almost any material can be made amorphous by rapid solidification from the melt (molten state). This condition is unstable, and the solid will crystallize in time. If the timescale for crystallization is years, then the amorphous state appears stable. Glasses are an example of amorphous solids. In crystalline silicon (Si) each atom is tetrahedrally bonded to four neighbours. In amorphous silicon (a-Si) the same short-range order exists, but the bond directions become changed at distances farther away from any atom. Amorphous silicon is a type of glass. Quasicrystals are another type of solid that lack long-range order.

Most solid materials found in nature exist in polycrystalline form rather than as a single crystal. They are actually composed of millions of grains (small crystals) packed together to fill all space. Each individual grain has a different orientation than its neighbours. Although long-range order exists within one grain, at the boundary between grains, the ordering changes direction. A typical piece of iron or copper (Cu) is polycrystalline. Single crystals of metals are soft and malleable, while polycrystalline metals are harder and stronger and are more useful industrially. Most polycrystalline materials can be made into large single crystals after extended heat treatment. In the past blacksmiths would heat a piece of metal to make it malleable: heat makes a few grains grow large by incorporating smaller ones. The smiths would bend the softened metal into shape and then pound it awhile; the pounding would make it polycrystalline again, increasing its strength.

Categories of Crystals

Crystals are classified in general categories, such as insulators, metals, semiconductors, and molecular solids. A single crystal of an insulator is usually transparent and resembles a piece of glass. Metals are shiny unless they have rusted. Semiconductors are sometimes shiny and sometimes transparent but are never rusty. Many crystals can be classified as a single type of solid, while others have intermediate behaviour. Cadmium sulfide (CdS) can be prepared in pure form and is an excellent insulator; when impurities are added to cadmium sulfide, it becomes an interesting semiconductor. Bismuth (Bi) appears to be a metal, but the number of electrons available for electrical conduction is similar to that of semiconductors. In fact, bismuth is called a semimetal. Molecular solids are usually crystals formed from molecules or polymers. They can be insulating, semiconducting, or metallic, depending on the type of molecules in the crystal. New molecules are continuously being synthesized, and many are made into crystals. The number of different crystals is enormous.

Structure

Crystals can be grown under moderate conditions from all 92 naturally occurring elements except helium, and helium can be crystallized at low temperatures by using 25 atmospheres of pressure. Binary crystals are composed of two elements. There are thousands of binary crystals; some examples are sodium chloride (NaCl), alumina (Al_2O_3), and ice (H_2O). Crystals can also be formed with three or more elements.

The Unit Cell

A basic concept in crystal structures is the unit cell. It is the smallest unit of volume that permits identical cells to be stacked together to fill all space. By repeating the pattern of the unit cell over and over in all directions, the entire crystal lattice can be constructed. A cube is the simplest example of a unit cell. Two other examples are shown in figure. The first is the unit cell for a face-centred cubic lattice, and the second is for a body-centred cubic lattice. These structures are explained in the following paragraphs. There are only a few different unit-cell shapes, so many different crystals share a single unit-cell type. An important characteristic of a unit cell is the number of atoms it contains. The total number of atoms in the entire crystal is the number in each cell multiplied by the number of unit cells. Copper and aluminum (Al) each have one atom per unit cell, while zinc (Zn) and sodium chloride have two. Most crystals have only a few atoms per unit cell, but there are some exceptions. Crystals of polymers, for example, have thousands of atoms in each unit cell.

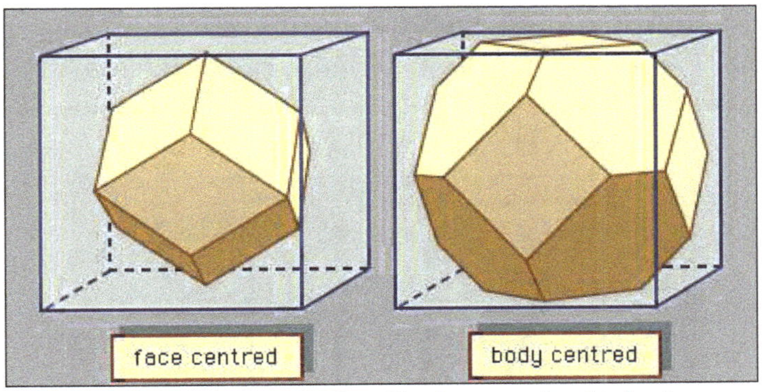

Unit cells for face-centred and body-centred cubic lattices.

Structures of Metals

The elements are found in a variety of crystal packing arrangements. The most common lattice structures for metals are those obtained by stacking the atomic spheres into the most compact arrangement. There are two such possible periodic arrangements. In each, the first layer has the atoms packed into a plane-triangular lattice in which every atom has six immediate neighbours. Figure shows this arrangement for the atoms labeled *A*. The second layer is shaded in the figure below. It has the same plane-triangular structure; the atoms sit in the holes formed by the first layer. The first layer has two equivalent sets of holes, but the atoms of the second layer can occupy only one set. The third layer, labeled *C*, has the same structure, but there are two choices for selecting the holes that the atoms will occupy. The third layer can be placed over the atoms of the first layer, generating an alternate layer sequence *ABABAB* . . ., which is called the hexagonal- closest-packed (hcp) structure. Cadmium and zinc crystallize with this structure. The second possibility is to place the atoms of the third layer over those of neither of the first two but instead over the set of holes in the first layer that remains unoccupied. The fourth layer is placed over the first, and so there is a three-layer repetition *ABCABCABC* . . ., which is called the face-centred cubic (fcc), or cubic-closest-packed, lattice. Copper, silver (Ag), and gold (Au) crystallize in fcc lattices. In the hcp and the fcc structures the spheres fill 74 percent of the volume, which represents the closest possible packing of spheres. Each atom has 12 neighbours. The number of atoms in a unit cell is two for hcp structures and one for fcc. There are 32 metals that have the hcp lattice and 26 with the fcc. Another possible arrangement is the body-centred cubic (bcc) lattice, in which each atom has eight neighbours arranged at the corners of a cube. Figure shows the cesium chloride (CsCl) structure, which is a cubic arrangement. If all atoms in this structure are of the same species, it is a bcc lattice. The spheres occupy 68 percent of the volume. There are 23 metals with the bcc arrangement. The sum of these three numbers (32 + 26 + 23) exceeds the number of elements that form metals (63), since some elements are found in two or three of these structures.

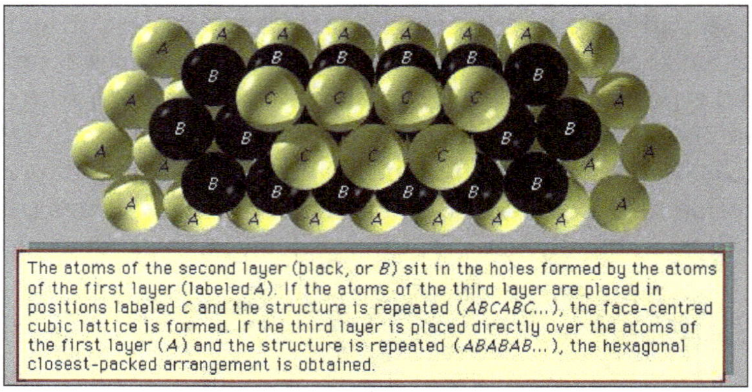

Stacking of spheres in closest-packed arrangements.

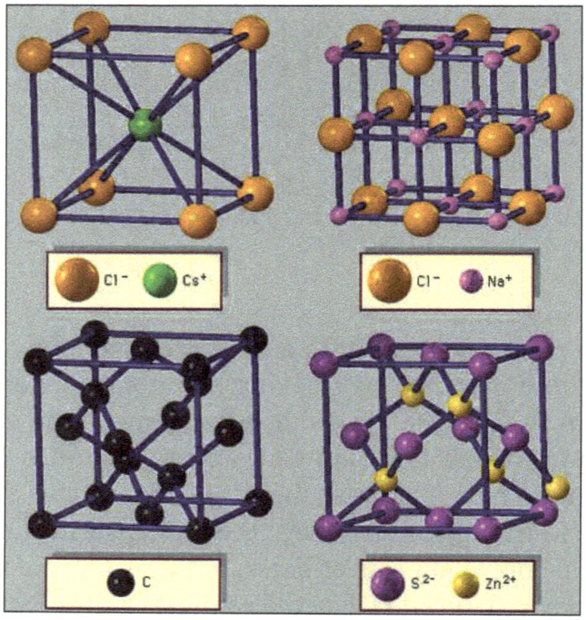

Crystal structures. There is an equal number of the two types of ions in the unit cell of the (A) cesium chloride, (B) sodium chloride, and (D) zinc blende arrangements. The diamond arrangement is shown in (C). If both atoms are identical in (A), the structure is body-centred cubic.

The fcc structure is also found for crystals of the rare gas solids neon (Ne), argon (Ar), krypton (Kr), and xenon (Xe). Their melting temperatures at atmospheric pressure are: Ne, 24.6 K; Ar, 83.8 K; Kr, 115.8 K; and Xe, 161.4 K.

Structures of Non-metallic Elements

The elements in the fourth row of the periodic table—carbon, silicon, germanium (Ge), and α-tin (α-Sn)—prefer covalent bonding. Carbon has several possible crystal structures. Each atom in the covalent bond has four first-neighbours, which are at the corners of a tetrahedron. This arrangement is called the diamond lattice and is shown in figure. There are two atoms in a unit cell, which is fcc. Large crystals of diamond are valuable gemstones. The crystal has other interesting properties; it has the highest sound velocity of any solid and is the best conductor of heat. Besides diamond, the other common form of carbon is graphite, which is a layered material. Each carbon atom has three coplanar near neighbours, forming an arrangement called the honeycomb lattice. Three-dimensional graphite crystals are obtained by stacking similar layers.

Another form of crystalline carbon is based on a molecule with 60 carbon atoms called buckminsterfullerene (C_{60}). The molecular shape is spherical. Each carbon is bonded to three neighbours, as in graphite, and the spherical shape is achieved by a mixture of 12 rings with five sides and 20 rings with six sides. Similar structures were first visualized by the American architect R. Buckminster Fuller for geodesic domes. The C_{60} molecules, also called buckyballs, are quite strong and almost incompressible. Crystals are formed such that the balls are arranged in an fcc lattice with a one-nanometre spacing between the centres of adjacent balls. The similar C_{70} molecule has the shape of a rugby ball; C_{70} molecules also form an fcc crystal when stacked together. The solid fullerenes form molecular crystals, with weak binding—provided by van der Waals interactions—between the molecules.

Many elements form diatomic gases: hydrogen (H), oxygen (O), nitrogen (N), fluorine (F), chlorine (Cl), bromine (Br), and iodine (I). When cooled to low temperature, they form solids of diatomic molecules. Nitrogen has the hcp structure, while oxygen has a more complex structure.

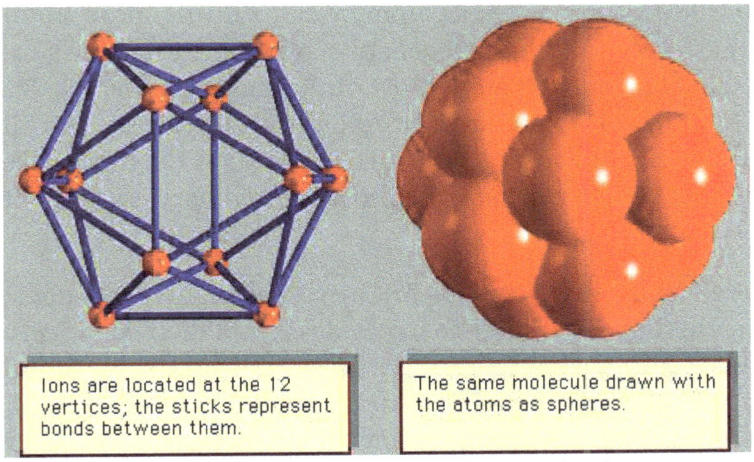

The icosahedral arrangement of boron molecules.

The most interesting crystal structures are those of elements that are neither metallic, covalent, nor diatomic. Although boron (B) and sulfur (S) have several different crystal structures, each has one arrangement in which it is usually found. Twelve boron atoms form a molecule in the shape of an icosahedron. Crystals are formed by stacking the molecules. The β-rhombohedral structure of boron has seven of these icosahedral molecules in each unit cell, giving a total of 84 atoms. Molecules of sulfur are usually arranged in rings; the most common ring has eight atoms.

The typical structure is α-sulfur, which has 16 molecules per unit cell, or 128 atoms. In the common crystals of selenium (Se) and tellurium (Te), the atoms are arranged in helical chains, which stack like cordwood. However, selenium also makes eight-atom rings, similar to sulfur, and forms crystals from them. Sulfur also makes helical chains, similar to selenium, and stacks them together into crystals.

Structures of Binary Crystals

Binary crystals are found in many structures. Some pairs of elements form more than one structure. At room temperature, cadmium sulfide may crystallize either in the zinc blende or wurtzite structure. Alumina also has two possible structures at room temperature, α-alumina (corundum) and β-alumina. Other binary crystals exhibit different structures at different temperatures. Among the most complex crystals are those of silicon dioxide (SiO_2), which has seven different structures at various temperatures and pressures; the most common of these structures is quartz. Some pairs of elements form several different crystals in which the ions have different chemical valences. Cadmium (Cd) and phosphorus (P) form the crystals Cd_3P_2, CdP_2, CdP_4, Cd_7P_{10}, and Cd_6P_7. Only in the first case are the ions assigned the expected chemical valences of Cd^{2+} and P^{3-}.

Among the binary crystals, the easiest structures to visualize are those with equal numbers of the two types of atoms. The structure of sodium chloride is based on a cube. To construct the lattice, the sodium and chlorine atoms are placed on alternate corners of a cube, and the structure is repeated. The structure of the sodium atoms alone, or the chlorine atoms alone, is fcc and defines the unit cell. The sodium chloride structure thus is made up of two interpenetrating fcc lattices. The cesium chloride lattice is based on the bcc structure; every other atom is cesium or chlorine. In this case, the unit cell is a cube. The third important structure for *AB* (binary) lattices is zinc blende. It is based on the diamond structure, where every other atom is *A* or *B*. Many binary semiconductors have this structure, including those with one atom from the third (boron, aluminum, gallium [Ga], or indium [In]) and one from the fifth (nitrogen, phosphorus, arsenic [As], or antimony [Sb]) column of the periodic table (GaAs, InP, etc.). Most of the chalcogenides (O, S, Se, Te) of cadmium and zinc (CdTe, ZnSe, ZnTe, etc.) also have the zinc blende structure. The mineral zinc blende is ZnS; its unit cell is also fcc. The wurtzite structure is based on the hcp lattice, where every other atom is *A* or *B*. These four structures comprise most of the binary crystals with equal numbers of cations and anions.

The fullerene molecule forms binary crystals M_xC_{60} with alkali atoms, where M is potassium (K), rubidium (Rb), or cesium (Cs). The fullerene molecules retain their spherical shape, and the alkali atoms sit between them. The subscript x can take on several values. A compound with $x = 6$ (*e.g.*, K_6C_{60}) is an insulator with the fullerenes in a bcc structure. The case $x = 4$ is an insulator with the body-centred tetragonal structure, while the case $x = 3$ is a metal with the fullerenes in an fcc structure. K_3C_{60}, Rb_3C_{60}, and Cs_3C_{60} are superconductors at low temperatures.

Alloys

Alloys are solid mixtures of atoms with metallic properties. The definition includes both amorphous and crystalline solids. Although many pairs of elements will mix together as solids, many pairs will not. Almost all chemical entities can be mixed in liquid form. But cooling a liquid to form a solid often results in phase separation; a polycrystalline material is obtained in which each grain is purely one atom or the other. Extremely rapid cooling can produce an amorphous alloy. Some pairs of elements form alloys that are metallic crystals. They have useful properties that differ from those exhibited by the pure elements. For example, alloying makes a metal stronger; for this reason alloys of gold, rather than the pure metal, are used in jewellery.

Atoms tend to form crystalline alloys when they are of similar size. The sizes of atoms are not easy to define, however, because atoms are not rigid objects with sharp boundaries. The outer part of an atom is composed of electrons in bound orbits; the average number of electrons decreases gradually with increasing distance from the nucleus. There is no point that can be assigned as the precise radius of the atom. Scientists have discovered, however, that each atom in a solid has a characteristic radius that determines its preferred separation from neighbouring atoms. For most types of atom this radius is constant, even in different solids. An empirical radius is assigned to each atom for bonding considerations, which leads to the concept of atomic size. Atoms readily make crystalline alloys when the radii of the two types of atoms agree to within roughly 15 percent.

Two kinds of ordering are found in crystalline alloys. Most alloys at low temperature are binary crystals with perfect ordering. An example is the alloy of copper and zinc. Copper is fcc, whereas zinc is hcp. A 50-percent-zinc–50-percent-copper alloy has a different structure—β-brass. At low temperatures it has the cesium chloride structure: a bcc lattice with alternating atoms of copper and zinc and a cubic unit cell. If the temperature is raised above 470° C, however, a phase transition to another crystalline state occurs. The ordering at high temperature is also bcc, but now each site has equal probability of having a copper or zinc atom. The two types of atoms randomly occupy each site, but there is still long-range order. At all temperatures, thousands of atoms away from a site, the location of the atom site can be predicted with certainty. At temperatures below 470° C one also knows whether that site will be occupied by a copper or zinc atom, while above 470° C there is an equal likelihood of finding either atom. The high-temperature phase is crystalline but disordered. The disorder phase is obtained through a partial melting, not into a liquid state but into a less ordered one. This behaviour is typical of metal alloys. Other common alloys are steel, an alloy of iron and carbon; stainless steel, an alloy of iron, nickel (Ni), and chromium (Cr); pewter and solder, alloys of tin and lead (Pb); and britannia metal, an alloy of tin, antimony, and copper.

Determination of Crystal Structures

Crystal structures are determined by scattering experiments using a portion of the crystal as the target. A beam of particles is sent toward the target, and upon impact some of the particles scatter from the crystal and ricochet in various directions. A measurement of the scattered particles provides raw data, which is then computer-processed to give a picture of the atomic arrangements. The positions are then inferred from the computer-analyzed data.

Max von Laue first suggested in 1912 that this measurement could be done using X rays, which are electromagnetic radiation of very high frequency. High frequencies are needed because these waves have a short wavelength. Von Laue realized that atoms have a spacing of only a few angstroms (1 angstrom [Å] is 10^{-10} metre, or 3.94×10^{-9} inch). In order to measure atomic arrangements, the particles scattering from the target must also have a wavelength of a few angstroms. X rays are required when the beam consists of electromagnetic radiation. The X rays only scatter in certain directions, and there are many X rays associated with each direction. The scattered particles appear in spots corresponding to locations where the scattering from each identical atom produces an outgoing wave that has all the wavelengths in phase. Figure shows incoming waves in phase. The scattering from atom A_2 has a longer path than that from atom A_1. If this additional path has a length (*AB* + *BC*) that is an exact multiple of the wavelength, then the two outgoing waves are in phase and reinforce each other. If the scattering angle is changed slightly, the waves no longer add coherently and begin to cancel one another. Combining the scattered radiation from all the atoms in the crystal causes all the outgoing waves to add coherently in certain directions and produce a strong signal in the scattered wave. If the extra path length (*AB* + *BC*) is five wavelengths, for example, the spot appears in one place. If it is six wavelengths, the spot is elsewhere. Thus, the different spots correspond to the different multiples of the wavelength of the X ray. The measurement produces two types of information: the directions of the spots and their intensity. This information is insufficient to deduce the exact crystal structure, however, as there is no algorithm by which the computer can go directly from the data to the structure. The crystallographer must propose various structures and compute how they would scatter the X rays. The theoretical results are compared with the measured one, and the theoretical arrangement is chosen that best fits the data. Although this procedure is fast when there are only a few atoms in a unit cell, it may take months or years for complex structures. Some protein molecules, for instance, have hundreds of atoms. Crystals of the proteins are grown, and X rays are used to measure the structure. The goal is to determine how the atoms are arranged in the protein, rather than how the proteins are arranged in the crystal.

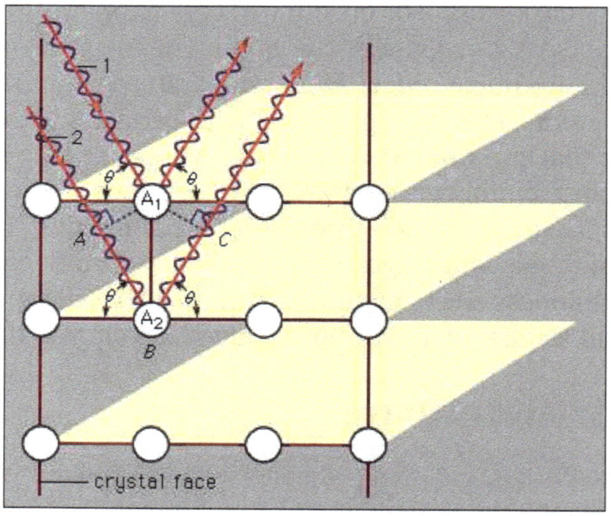

Incident rays (1 and 2) at angle θ on the planes of atoms in a crystal. Rays reinforce if their difference in path length (*AB* + *BC*) is an integer times the wavelength of the X ray.

Beams of neutrons may also be used to measure crystal structure. The beam of neutrons is obtained by drilling a hole in the side of a nuclear reactor. The energetic neutrons created in nuclear fission escape through the hole. The motion of elementary particles is governed by quantum, or wave, mechanics. Each neutron has a wavelength that depends on its momentum. The scattering directions are determined by the wavelength, as is the case with X rays. The wavelengths for neutrons from a reactor are suitable for measuring crystal structures.

X rays and neutrons provide the basis for two competing technologies in crystallography. Although they are similar in principle, the two methods have some differences. X rays scatter from the electrons in the atoms so that more electrons result in more scattering. X rays easily detect atoms of high atomic number, which have many electrons, but cannot readily locate atoms with few electrons. In hydrogen-bonded crystals, X rays do not detect the protons at all. Neutrons, on the other hand, scatter from the atomic nucleus. They scatter readily from protons and are excellent for determining the structure of hydrogen-bonded solids. One drawback to this method is that some nuclei absorb neutrons completely, and there is little scattering from these targets.

Beams of electrons can also be used to measure crystal structure, because energetic electrons have a wavelength that is suitable for such measurements. The problem with electrons is that they scatter strongly from atoms. Proper interpretation of the experimental results requires that an electron scatter only from one atom and leave the crystal without scattering again. Low-energy electrons scatter many times, and the interpretation must reflect this. Low-energy electron diffraction (LEED) is a technique in which a beam of electrons is directed toward the surface. The scattered electrons that reflect backward from the surface are measured. They scatter many times before leaving backward but mainly leave in a few directions that appear as "spots" in the measurements. An analysis of the varied spots gives information on the crystalline arrangement. Because the electrons are scattered strongly by the atoms in the first few layers of the surface, the measurement gives only the arrangements of atoms in these layers. It is assumed that the same structure is repeated throughout the crystal. Another scattering experiment involves electrons of extremely high energy. The scattering rate decreases as the energy of the electron increases, so that very energetic electrons usually scatter only once. Various electron microscopes are constructed on this principle.

Types of Bonds

The properties of a solid can usually be predicted from the valence and bonding preferences of its constituent atoms. Four main bonding types are discussed here: ionic, covalent, metallic, and molecular. Hydrogen-bonded solids, such as ice, make up another category that is important in a few crystals. There are many examples of solids that have a single bonding type, while other solids have a mixture of types, such as covalent and metallic or covalent and ionic.

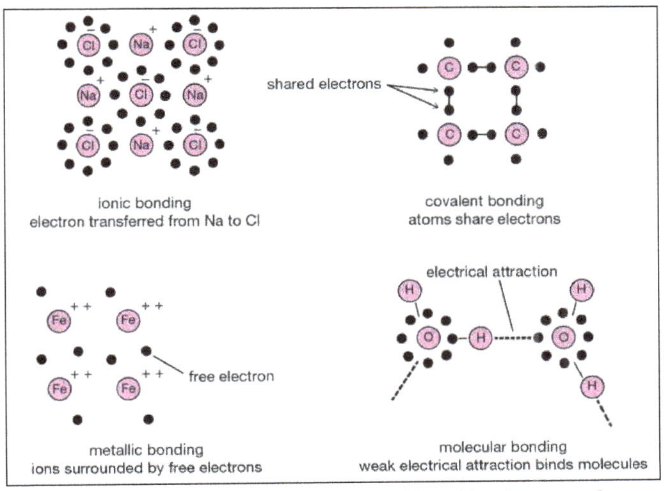

Crystal Bonding Different types of bonding in crystals.

Ionic Bonds

Sodium chloride exhibits ionic bonding. The sodium atom has a single electron in its outermost shell, while chlorine needs one electron to fill its outer shell. Sodium donates one electron to chlorine, forming a sodium ion (Na^+) and a chlorine ion (Cl^-). Each ion thus attains a closed outer shell of electrons and takes on a spherical shape. In addition to having filled shells and a spherical shape, the ions of an ionic solid have integer valence. An ion with positive valence is called a cation. In an ionic solid the cations are surrounded by ions with negative valence, called anions. Similarly, each anion is surrounded by cations. Since opposite charges attract, the preferred bonding occurs when each ion has as many neighbours as possible, consistent with the ion radii. Six or eight nearest neighbours are typical; the number depends on the size of the ions and not on the bond angles. The alkali halide crystals are binaries of the *AH* type, where *A* is an alkali ion (lithium [Li], sodium, potassium, rubidium, or cesium) and *H* is a halide ion (fluorine, chlorine, bromine, or iodine). The crystals have ionic bonding, and each ion has six or eight neighbours. Metal ions in the alkaline earth series (magnesium [Mg], calcium [Ca], barium [Ba], and strontium [Sr]) have two electrons in their outer shells and form divalent cations in ionic crystals. The chalcogenides (oxygen, sulfur, selenium, and tellurium) need two electrons to fill their outer *p*-shell. (Electron shells are divided into subshells, designated as *s*, *p*, *d*, *f*, *g*, and so forth. Each subshell is divided further into orbitals.) Two electrons are transferred from the cations to the anions, leaving each with a closed shell. The alkaline earth chalcogenides form ionic binary crystals such as barium oxide (BaO), calcium sulfide (CaS), barium selenide (BaSe), or strontium oxide (SrO). They have the same structure as sodium chloride, with each atom having six neighbours. Oxygen can be combined with various cations to form a large number of ionically bonded solids.

Covalent Bonds

Silicon, carbon, germanium, and a few other elements form covalently bonded solids. In these elements there are four electrons in the outer *sp*-shell, which is half filled. (The *sp*-shell is a hybrid formed from one *s* and one *p* subshell.) In the covalent bond an atom shares one valence (outer-shell) electron with each of its four nearest neighbour atoms. The bonds are highly directional and prefer a tetrahedral arrangement. A covalent bond is formed by two electrons—one from each atom—located in orbitals between the ions. Insulators, in contrast, have all their electrons within shells inside the atoms.

The perpetual spin of an electron is an important aspect of the covalent bond. From a vantage point above the spinning particle, counterclockwise rotation is designated spin-up, while clockwise rotation is spin-down. A fundamental law of quantum physics is the Pauli exclusion principle, which states that no two electrons can occupy the same point in space at the same time with the same direction of spin. In a covalent bond two electrons occupy the same small volume of space (*i.e.*, the same orbital) at all times, so they must have opposite spin: one up and one down. The exclusion principle is then satisfied, and the resulting bond is strong.

In graphite the carbon atoms are arranged in parallel sheets, and each atom has only three near neighbours. The covalent bonds between adjacent carbons within each layer are quite strong and are called σ bonds. The fourth valence electron in carbon has its orbital perpendicular to the plane. This orbital bonds weakly with the similar orbitals on all three neighbours, forming π bonds. The four bonds for each carbon atom in the graphite structure are not arranged in a tetrahedron; three are in a plane. The planar arrangement results in strong bonding, although not as strong as the bonding in the diamond configuration. The bonding between layers is quite weak and arises from the van der Waals interaction; there is much slippage parallel to the layers. Diamond and graphite form an interesting contrast: diamond is the hardest material in nature and is used as an abrasive, while graphite is used as a lubricant.

Besides the elemental semiconductors, such as silicon and germanium, some binary crystals are covalently bonded. Gallium has three electrons in the outer shell, while arsenic lacks three. Gallium arsenide (GaAs) could be formed as an insulator by transferring three electrons from gallium to arsenic; however, this does not occur. Instead, the bonding is more covalent, and gallium arsenide is a covalent semiconductor. The outer shells of the gallium atoms contribute three electrons, and those of the arsenic atoms contribute five, providing the eight electrons needed for four covalent bonds. The centres of the bonds are not at the midpoint between the ions but are shifted slightly toward the arsenic. Such bonding is typical of the III–V semiconductors—*i.e.*, those consisting of one element from the third column of the periodic table and one from the fifth column. Elements from the third column (boron, aluminum, gallium, and indium) contribute three electrons, while the fifth-column elements (nitrogen, phosphorus, arsenic, and antimony) contribute five electrons. All III–V semiconductors are covalently bonded and typically have the zinc blende structure with four neighbours per atom. Most common semiconductors favour this arrangement.

The factor that determines whether a binary crystal will act as an insulator or a semiconductor is the valence of its constituent atoms. Ions that donate or accept one or two valence electrons form insulators. Those that have three to five valence electrons tend to have covalent bonds and form semiconductors. There are exceptions to these rules, however, as is the case with the IV–VI semiconductors such as lead sulfide. Heavier elements from the fourth column of the periodic table (germanium, tin, and lead) combine with the chalcogenides from the sixth row to form good binary semiconductors such as germanium telluride (GeTe) or tin sulfide (SnS). They have the sodium chloride structure, where each atom has six neighbours. Although not tetrahedrally bonded, they are good semiconductors.

Filled atomic shells with *d*-orbitals have an important role in covalent bonding. Electrons in atomic orbits have angular momentum (L), which is quantized in integer (n) multiples of Planck's constant h: $L = nh$. Electron orbitals with $n = 0$ are called *s*-states, with $n = 1$ are *p*-states, and with $n = 2$ are *d*-states. Silver and copper ions have one valence electron outside their closed shells. The outermost filled shell is a *d*-state and affects the bonding. Eight binary crystals are formed from the copper and silver halides. Three (AgF, AgCl, AgBr) have the sodium chloride structure with six neighbours. The other five (AgI, CuF, CuCl, CuBr, CuI) have the zinc blende structure with four neighbours. The bonding in this group of solids is on the borderline between covalent and ionic, since the crystals prefer both types of bonds. The alkali metal halides exhibit somewhat different behaviour. The alkali metals are also monovalent cations, but their halides are strictly ionic. The difference in bonding between the alkali metals on the one hand and silver and copper on the other hand is that silver and copper have filled *d*-shells while the alkalis have filled *p*-shells. Since the *d*-shells are filled, they do not covalently bond. This group of electrons is, however, highly polarizable, which influences the bonding of the valence electrons. Similar behaviour is found for zinc and cadmium, which have two valence electrons outside a filled *d*-shell. They form binary crystals with the chalcogenides, which have tetrahedral bonding. In this case the covalent bonding seems to be preferred over the ionic bond. In contrast, the alkaline earth chalcogenides, which are also divalent, have outer *p*-shells and are ionic. The zinc and cadmium chalcogenides are covalent, as the outer *d*-shell electrons of the two cations favour covalent bonding.

Metallic Bonds

Metallic bonds fall into two categories. The first is the case in which the valence electrons are from the *sp*-shells of the metal ions; this bonding is quite weak. In the second category the valence electrons are from partially filled *d*-shells, and this bonding is quite strong. The *d*-bonds dominate when both types of bonding are present.

The simple metals are bonded with *sp*-electrons. The electrons of these metal atoms are in filled atomic shells except for a few electrons that are in unfilled *sp*-shells. The electrons from the unfilled shells are detached from the metal ion and are free to wander throughout the crystal. They are called conduction electrons, since they are responsible for the electrical conductivity of metals. Although the conduction electrons may roam anywhere in the crystal, they are distributed uniformly throughout the entire solid. Any large imbalance of charge is prevented by the strong electrical attraction between the negative electrons and the positive ions, plus the strong repulsion between electrons. The phrase electron correlation describes the correlated movements of the electrons; the motion of each electron depends on the positions of neighbouring electrons. Electrons have strong short-range order with one another. Correlation ensures that each unit cell in the crystal has, on the average, the number of electrons needed to cancel the positive charge of the cation so that the unit cell is electrically neutral.

Cohesive energy is the energy gained by arranging the atoms in a crystalline state, as compared with the gas state. Insulators and semiconductors have large cohesive energies; these solids are bound together strongly and have good mechanical strength. Metals with electrons in *sp*-bonds have very small cohesive energies. This type of metallic bond is weak; the crystals are barely held together. Single crystals of simple metals such as sodium are mechanically weak. At room temperature the crystals have the mechanical consistency of warm butter. Special care must be used in handling these crystals, because they are easily distorted. Metals such as magnesium or aluminum must be alloyed or polycrystalline to have any mechanical strength. Although the simple metals are found in a variety of structures, most are in one of the three closest-packed structures: fcc, bcc, and hcp. Theoretical calculations show that the cohesive energy of a given metal is almost the same in each of the different crystal arrangements; therefore, crystal arrangements are unimportant in metals bound with electrons from *sp*-shells.

A different type of metallic bonding is found in transition metals, which are metals whose atoms are characterized by unfilled *d*-shells. The *d*-orbitals are more tightly bound to an ion than the *sp*-orbitals. Electrons in *d*-shells do not wander away from the ion. The *d*-orbitals form a covalent bond with the *d*-orbitals on the neighbouring atoms. The bonding of *d*-orbitals does not occur in a tetrahedral arrangement but has a different directional preference. In metals the bonds from *d*-orbitals are not completely filled with electrons. This situation is different from the tetrahedral bonds in semiconductors, which are filled with eight electrons. In transition metals the covalent bonds formed with the *d*-electrons are much stronger than the weak bonds made with the *sp*-electrons of simple metals. The cohesive energy is much larger in transition metals. Titanium, iron, and tungsten, for example, have exceptional mechanical strength. Crystal arrangements are important in the behaviour of the transition metals and occur in the close-packed fcc, bcc, or hcp arrangements.

Molecular Binding

The Dutch physicist Johannes D. van der Waals first proposed the force that binds molecular solids. Any two atoms or molecules have a force of attraction (*F*) that varies according to the inverse seventh power of the distance *R* between the centres of the atoms or molecules: $F = -C/R^7$, where C is a constant. The force, known as the van der Waals force, declines rapidly with the distance *R* and is quite weak. If the atoms or molecules have a net charge, there is a strong force whose strength varies according to Coulomb's law as the inverse second power of the separation distance: $F = -C'/R^2$, where C′ is a constant. This force provides the binding in ionic crystals and some of the binding in metals. Coulomb's law does not apply to atoms or molecules without a net charge. Molecules with a dipole moment, such as water, have a strong attractive force owing to the interactions between the dipoles. For atoms and molecules with neither net charges nor dipole moments, the van der Waals force provides the crystal binding. The force of gravity also acts between neutral atoms and molecules, but it is far too weak to bind molecules into crystals.

The van der Waals force is caused by quantum fluctuations. Two neighbouring atoms that are each fluctuating can lower their joint energy by correlating their fluctuations. The van der Waals force arises from correlations in their dipole fluctuations. Electrons bound in atomic orbits are in constant motion around the nucleus, and the distribution of charges in the atom changes constantly as the electrons move, owing to quantum fluctuations. One fluctuation might produce a momentary electric dipole moment (*i.e.*, a separation of charges) on an atom if a majority of its electrons are on one side of the nucleus. The dipole moment creates an electric field on a neighbouring atom; this field will induce a dipole moment on the second atom. The two dipoles attract one another via the van der Waals interaction. Since the force depends on the inverse seventh power of the distance, it declines rapidly with increasing distance. Atoms have a typical radius of one to three angstroms. The van der Waals force binds atoms and molecules within a few angstroms of each other; beyond that distance the force is negligible. Although weak, the van der Waals force is always present and is important in cases where the other forces are absent.

Hydrogen is rarely found as a single atom. Instead it forms diatomic molecules (H_2), which are gaseous at room temperature. At lower temperatures the hydrogen becomes a liquid and at about 20 K turns into a solid. The molecule retains its identity in the liquid and solid states. The solid exists as a molecular crystal of covalently bound H_2 molecules. The molecules attract one another by van der Waals forces, which provide the crystal binding. Helium, the second element in the periodic table, has two electrons, which constitute a filled atomic shell. In its liquid and solid states, the helium atoms are bound together by van der Waals forces. In fact, all the rare gases (helium, neon, argon, krypton, and xenon) are molecular crystals with the binding provided by van der Waals forces.

Many organic molecules form crystals where the molecules are bound by van der Waals forces. In methane (CH_4), a central carbon makes a covalent bond with each hydrogen atom, forming a tetrahedron. In crystalline methane the molecules are arranged in the fcc structure. Benzene (C_6H_6) has the carbon atoms in a hexagonal ring; each carbon has three coplanar σ bonds, as in graphite, where two bonds are with neighbouring carbon atoms and the third bond is with a hydrogen atom. Crystalline benzene has four molecules per unit cell in a complex arrangement. Fullerene and the rare gas atoms are spherical, and the crystalline arrangement corresponds to the closest packing of spheres. Most organic molecules, however, are not spherical and display irregular shapes. For odd-shaped molecules, the van der Waals interaction depends on the rotational orientation of the two molecules. In order to maximize the force, the molecules in the crystal have unusual arrangements, as in the case of benzene.

Hydrogen Bonding

Hydrogen bonding is important in a few crystals, notably in ice. With its lone electron, a hydrogen atom usually forms a single covalent bond with an electronegative atom. In the hydrogen bond the atom is ionized to a proton. The proton sits between two anions and joins them. Hydrogen bonding occurs with only the most electronegative ions: nitrogen, oxygen, and fluorine. In water the hydrogen links pairs of oxygen ions. Water is found in many different crystal structures, but they all have the feature that the hydrogen atoms sit between pairs of oxygen. Another hydrogen-bonded solid is hydrogen fluoride (HF), in which the hydrogen atom (proton) links pairs of fluorines.

Crystal Growth

The earliest crystal grower was nature. Many excellent crystals of minerals formed in the geologic past are found in mines and caves throughout the world. Most precious and semiprecious stones are well-formed crystals. Early efforts to produce synthetic crystals were concentrated on making gems. Synthetic ruby was grown by the French scientist Marc Antoine Augustin Gaudin in 1873. Since about 1950 scientists have learned to grow in the laboratory crystals of quality equal or superior to those found in nature. New techniques for growth are continually being developed, and crystals with three or more atoms per unit cell are continually being discovered.

Dunite and large crystals of selenite.

Mineral crystals vary greatly in size. The photo on the left shows an extremely thin slice of dunite, an igneous rock, as viewed with a special microscope. Each different-coloured shape in this slice is one of the many minerals that make up the rock; each mineral is only a few millimetres or less across. The photo on the right shows extremely large crystals made of selenite, many of which are several metres long.

Vapour Growth

Crystals can be grown from a vapour when the molecules of the gas attach themselves to a surface and move into the crystal arrangement. Several important conditions must be met for this to occur. At constant temperature and equilibrium conditions, the average number of molecules in the gas and solid states is constant; molecules leave the gas and attach to the surface at the same rate that they leave the surface to become gas molecules. For crystals to grow, the gas-solid chemical system must be in a nonequilibrium state such that there are too many gaseous molecules for the conditions of pressure and temperature. This state is called supersaturation. Molecules are more prone to leave the gas than to rejoin it, so they become deposited on the surface of the container. Supersaturation can be induced by maintaining the crystal at a lower temperature than the gas. A critical stage in the growth of a crystal is seeding, in which a small piece of crystal of the proper structure and orientation, called a seed, is introduced into the container. The gas molecules find the seed a more favourable surface than the walls and preferentially deposit there. Once the molecule is on the surface of the seed, it wanders around this surface to find the preferred site for attachment. Growth proceeds one molecule at a time and one layer at a time. The process is slow; it takes days to grow a small crystal. Crystals are grown at temperatures well below the melting point to reduce the density of defects. The advantage of vapour growth is that very pure crystals can be grown by this method, while the disadvantage is that it is slow.

Most clouds in the atmosphere are ice crystals that form by vapour growth from water molecules. Most raindrops are crystals as they begin descending but thaw during their fall to Earth. Seeding for rain—accomplished by dropping silver iodide crystals from airplanes—is known to induce precipitation. In the laboratory, vapour growth is usually accomplished by flowing a supersaturated gas over a seed crystal. Quite often a chemical reaction at the surface is needed to deposit the atoms. Crystals of silicon can be grown by flowing chlorosilane ($SiCl_4$) and hydrogen (H_2) over a seed crystal of silicon. Hydrogen acts as the buffer gas by controlling the temperature and rate of flow. The molecules dissociate on the surface in a chemical reaction that forms hydrogen chloride (HCl) molecules. Hydrogen chloride molecules leave the surface, while silicon atoms remain to grow into a crystal. Binary crystals such as gallium arsenide (GaAs) are grown by a similar method. One process employs gallium chloride (GaCl) as the gallium carrier. Arsenic is provided by molecules such as arsenous chloride ($AsCl_3$), arsine (AsH_3), or As_4 (yellow arsenic). These molecules, with hydrogen as the buffer gas, grow crystals of gallium arsenide while forming gas molecules such as gallium trichloride ($GaCl_3$) and hydrogen chloride. Trimethylgallium, $(CH_3)_3Ga$, is another molecule that can be used to deliver gallium to the surface.

Growth from Solution

Large single crystals may be grown from solution. In this technique the seed crystal is immersed in a solvent that contains typically about 10–30 percent of the desired solute. The choice of solvent usually depends on the solubility of the solute. The temperature and pH (a measure of acidity or basicity) of the solution must be well controlled. The method is faster than vapour growth, because there is a higher concentration of molecules at the surface in a liquid as compared to a gas, but it is still relatively slow.

Growth from the Melt

This method is the most basic. A gas is cooled until it becomes a liquid, which is then cooled further until it becomes a solid. Polycrystalline solids are typically produced by this method unless special techniques are employed. In any case, the temperature must be controlled carefully. Large crystals can be grown rapidly from the liquid elements using a popular method invented in 1918 by the Polish scientist Jan Czochralski and called crystal pulling. One attaches a seed crystal to the bottom of a vertical arm such that the seed is barely in contact with the material at the surface of the melt. A modern Czochralski apparatus is shown in figure. The arm is raised slowly, and a crystal grows underneath at the interface between the crystal and the melt. Usually the crystal is rotated slowly, so that inhomogeneities in the liquid are not replicated in the crystal. Large-diameter crystals of silicon are grown in this way for use as computer chips. Based on measurements of the weight of the crystal during the pulling process, computer-controlled apparatuses can vary the pulling rate to produce any desired diameter. Crystal pulling is the least expensive way to grow large amounts of pure crystal. A photograph of a single crystal of stainless steel grown by the Czochralski method is shown in figure. The original seed is on the right tip. Binary crystals can also be pulled; for example, synthetic sapphire crystals can be pulled from molten alumina. Special care is required to grow binary and other multicomponent crystals; the temperature must be precisely controlled because such crystals may be grown only at a single, extremely high temperature. The melt has a tendency to be inhomogeneous, since the two liquids may try to separate by gravity.

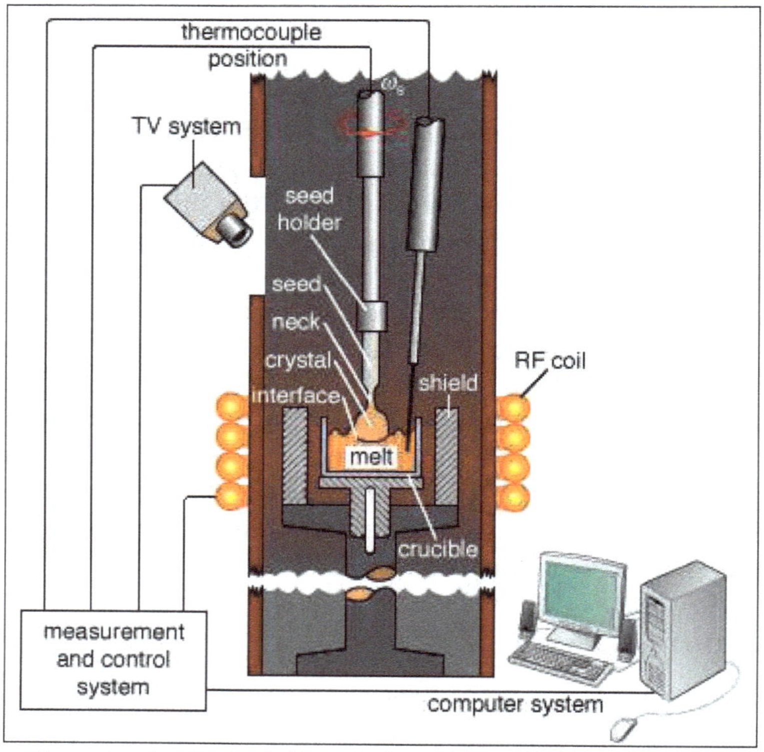

Crystal pulling using the Czochralski method. A schematic view of a modern apparatus.

The Bridgman method (named after the American scientist Percy Williams Bridgman) is also widely used for growing large single crystals. The molten material is put into a crucible, often of silica, which has a cylindrical shape with a conical lower end. Heaters maintain the molten state. As the crucible is slowly lowered into a cooler region, a crystal starts growing in the conical tip. The crucible is lowered at a rate that matches the growth of the crystal, so that the interface between crystal and melt is always at the same temperature. The rate of moving the crucible depends on the temperature and the material. When done successfully, the entire molten material in the crucible grows into a single large crystal. One disadvantage of the method is that excess impurities are pushed out of the crystal during growth. A layer of impurities grows at the interface between melt and solid as this surface moves up the melt, and the impurities become concentrated in the higher part of the crystal.

Epitaxy is the technique of growing a crystal, layer by layer, on the atomically flat surface of another crystal. In homoepitaxy a crystal is grown on a substrate of the same material. Silicon layers of different impurity content, for example, are grown on silicon substrates in the manufacture of computer chips. Heteroepitaxy, on the other hand, is the growth of one crystal on the substrate of another. Silicon substrates are often used since they are readily available in atomically smooth form. Many different semiconductor crystals can be grown on silicon, such as gallium arsenide, germanium, cadmium telluride (CdTe), and lead telluride (PbTe). Any flat substrate can be used for epitaxy, however, and insulators such as rock salt (NaCl) and magnesium oxide (MgO) are also used.

Molecular-beam epitaxy, commonly abbreviated as MBE, is a form of vapour growth. The field began when the American scientist John Read Arthur reported in 1968 that gallium arsenide could be grown by sending a beam of gallium atoms and arsenic molecules toward the flat surface of a crystal of the molecule. The amount of gas molecules can be controlled to grow just one layer, or just two, or any desired amount. This method is slow, since molecular beams have low densities of atoms. Chemical vapour deposition (CVD) is another form of epitaxy that makes use of the vapour growth technique. Also known as vapour-phase epitaxy (VPE), it is much faster than MBE since the atoms are delivered in a flowing gas rather than in a molecular beam. Synthetic diamonds are grown by CVD. Rapid growth occurs when methane (CH_4) is mixed with atomic hydrogen gas, which serves as a catalyst. Methane dissociates on a heated surface of diamond. The carbon remains on the surface, and the hydrogen leaves as a molecule. Growth rates are several micrometres (1 micrometre is equal to 0.00004 inch) per hour. At that rate, a stone 1 centimetre (0.4 inch) thick is grown in 18 weeks. CVD diamonds are of poor quality as gemstones but are important electronic materials. Because hydrogen is found in nature as a molecule rather than as a single atom, making atomic hydrogen gas is the major expense in growing CVD diamonds. Liquid-phase epitaxy (LPE) uses the solution method to grow crystals on a substrate. The substrate is placed in a solution with a saturated concentration of solute. This technique is used to grow many crystals employed in modern electronics and optoelectronic devices, such as gallium arsenide, gallium aluminum arsenide, and gallium phosphide.

An important concern in successful epitaxy is matching lattice distances. If the spacing between atoms in the substrate is close to that of the top crystal, then that crystal will grow well; a small difference in lattice distance can be accommodated as the top crystal grows. When the lattice distances are different, however, the top crystal becomes deformed, since structural defects such as dislocations appear. Although few crystals share the same lattice distance, a number of examples are known. Aluminum arsenide and gallium arsenide have the same crystal structure and the same lattice parameters to within 0.1 percent; they grow excellent crystals on one another. Such materials, known as superlattices, have a repeated structure of n layers of GaAs, m layers of AlAs, n layers of GaAs, m layers of AlAs, and so forth. Superlattices represent artificially created structures that are thermodynamically stable; they have many applications in the modern electronics industry. Another lattice-matched epitaxial system is mercury telluride (HgTe) and cadmium telluride (CdTe). These two semiconductors form a continuous semiconductor alloy $Cd_xHg_{1-x}Te$, where x is any number between 0 and 1. This alloy is used as a detector of infrared radiation and is incorporated in particular in night-vision goggles.

Dendritic Growth

At slow rates of crystal growth, the interface between melt and solid remains planar, and growth occurs uniformly across the surface. At faster rates of crystal growth, instabilities are more likely to occur; this leads to dendritic growth. Solidification releases excess energy in the form of heat at the interface between solid and melt. At slow growth rates, the heat leaves the surface by diffusion. Rapid growth creates more heat, which is dissipated by convection (liquid flow) when diffusion is too slow. Convection breaks the planar symmetry so that crystal growth develops along columns, or "fingers," rather than along planes. Each crystal has certain directions in which growth is fastest, and dendrites grow in these directions. As the columns grow larger, their surfaces become flatter and more unstable. This feather or tree structure is characteristic of dendritic growth. Snowflakes are an example of crystals that result from dendritic growth.

Electric Properties

Resistivity

The German physicist Georg Simon Ohm discovered the basic law of electric conduction, which is now called Ohm's law. His law relates the voltage (V, measured in volts), the current (I, in amperes), and the resistance (R, in ohms) according to the formula $V = RI$. A current I through a solid induces a voltage V; the resistance R is the constant of proportionality. The value of R is an important factor in the design of electrical circuits. It is determined by the shape of the resistor: a long narrow object has more resistance than a short wide one of the same material. For solids, the important parameter is the resistivity ρ, which is given in units of ohm-metres. It is the resistivity per volume unit and is independent of shape. The relationship between R and ρ is $R = \rho L/A$, where A is the area of the resistor and L is the length. These dimensions are measured in the direction of the current: L is the length of the current path, and A is the cross-sectional area. The resistance of a copper bar depends on its shape, but at a given temperature every piece of pure copper has the same resistivity. Thus the resistivity is a fundamental parameter of a material and is investigated by scientists. Resistivities of solids span a wide range of values. Certain metals have zero resistivity at low temperatures; they are called superconductors. At the other extreme, very good insulators such as sulfur and polystyrene have resistivities larger than one quadrillion ohm-metres. At room temperature, the metal with the lowest value of resistivity is silver, with $\rho = 1.6 \times 10^{-8}$ ohm-metre; the second best conductor is copper, with $\rho = 1.7 \times 10^{-8}$ ohm-metre. Copper, rather than silver, is used in household wires because of the high cost of silver.

Conduction through Ion Hopping

Electrical conductivity σ is the inverse of resistivity and is measured in units of ohm-metre^{-1}. Electrical current is produced by the motion of charges. In crystals, electrical current is due to the motion of both ions and electrons. Ions move by hopping occasionally from site to site; all solids can conduct electricity in this manner. When the voltage is zero, there is no net current because the ions hop randomly in all directions. The imposition of a small voltage causes the ions to slightly favour one direction of motion, which leads to a net flow of charge in that direction; this constitutes an electrical current. The electricity conducted by this process is quite small and is usually negligible compared with that carried by the electrons. When an ion hops, it must migrate to a vacant site, which could be either an interstitial or a vacancy. Ionic conductivity can occur because hopping ions cause vacancies to move through the solid. An ion hops to the vacant site, thereby filling the vacancy, while creating a new one at the ion's former site. Repeating this process causes the position of the vacancy to migrate through the crystal. The motion of the vacancy arises from the motion of ions, which carry charge and contribute to electrical conductivity.

Ion hops are induced by thermal fluctuations. Most of the ions move within their lattice site, vibrating around this point. Temperature is defined as the average energy of this vibrational motion; the more the ions move, the higher the temperature. An individual ion at times moves slowly and at times vibrates quite rapidly but usually has an energy near the average value. Each ion shares its vibrational energy with its neighbouring ions. An ion typically has some neighbours with small vibrations and others with large ones. The average energy shared with the neighbours is close to the average energy of all the atoms. As a random process, however, it occasionally happens that all neighbours of an ion may have large vibrations, in which case the ion will acquire an unusually high energy. This energy may be high enough to cause it to leave its site and hop to a neighbouring site. A thermal fluctuation is the rare process in which the energy at a local site may be much higher or lower than the average energy in the crystal. Probability theory shows that the higher the temperature, the more frequent are these thermal fluctuations. Ions therefore hop more often at high temperature.

A few solids conduct electricity better by ion motion than by electron motion. These unusual materials are technologically important in making batteries. All batteries have two electrodes separated by an electrolyte, which is a material that conducts ions better than electrons. An example of a crystal electrolyte is β-alumina, which readily conducts monovalent cations such as silver (Ag^+) and sodium (Na^+). Among all ions, silver has the largest value of ionic conductivity in many different electronic insulators. The copper ion (Cu^+) forms the same type of chemical bonds as does the silver ion, but the copper ion, because of its smaller radius, does not migrate as well within an electrolyte. Silver ions fit perfectly into the interstitial sites of the crystal lattices of several electrolytes, while the smaller copper ions permit the neighbouring ions to collapse around them, inhibiting further hopping. There are a few good conductors of the inexpensive copper ion that can be used as solid electrolytes in batteries. Silver is too costly and heavy to use in large-volume batteries such as those found in automobiles, but it is used in the smaller batteries that power devices such as hearing aids.

Conduction Electrons

Electrons carry the basic unit of charge e, equal to 1.6022×10^{-19} coulomb. They have a small mass and move rapidly. Most electrons in solids are bound to the atoms in local orbits, but a small fraction of the electrons are available to move easily through the entire crystal. These so-called conduction electrons carry the electrical current. Solids with many conduction electrons are metals, while those with a few are semimetals or semiconductors. In insulators, nearly all the electrons are bound, and very few electrons are capable of carrying current. A typical metal has one or more conduction electrons in each atomic unit cell, a semiconductor may have only one conduction electron for each thousand unit cells, and an insulator may have one conduction electron per one million or one trillion unit cells.

The bonding properties of the individual atoms of a solid determine the behaviour of the bulk solid. The electrical properties of a solid can usually be predicted from the valence and bonding preferences of its atoms. In the argon atom, for example, all atomic shells are filled with electrons. The electrons of solid argon remain in the atomic shells; none are conduction electrons, and the electrical resistivity is therefore high. Solid argon, like all the rare gas solids, is a good insulator. A few conduction electrons are contributed by impurities, and so the conductivity, though small, is not zero. These conduction electrons move quite readily through the solid. The term mobility is used to describe how well a conduction electron moves through the solid in response to a voltage. Conductivity is the product of mobility, the electrical charge e, and the number N of conduction electrons per unit volume: $\sigma = Ne\mu$, where σ is the conductivity and μ is the mobility. The mobility of the rare gas solids is high, but their conductivity is nonetheless low because there is a small number of conduction electrons.

Electrical Insulators

Like the rare gas solids, most ionic solids are electrical insulators. In sodium chloride, for example, each sodium atom donates its single valence electron to a chlorine atom, thus forming a solid composed of Na^+ and Cl^- ions. All electrons are in filled shells at low temperature, and in a perfect crystal there are no conduction electrons. Sodium chloride is thus an insulator with a very high resistivity. Some conduction electrons are provided by impurities or thermal excitations. At high temperatures large ion vibrations from thermal fluctuations may knock an electron out of a filled shell, upon which it becomes a conduction electron and contributes to the conductivity. The number of conduction electrons created by thermal excitations is small for most insulators. Although defects can be responsible for producing conduction electrons, they can also destroy the conducting ability of electrons by trapping them. The defects have local orbitals that provide a lower energy state for the electron than the one occupied in the conduction state. A conduction electron becomes bound at the defect, ceasing to contribute to the conductivity. This process is very efficient in insulators, so the few conduction electrons provided by impurities and thermal fluctuations are usually trapped at other defects. By definition, an insulator is a solid that does not provide a stable environment for conduction electrons.

Conductivity of Metals

Metals have a high density of conduction electrons. The aluminum atom has three valence electrons in a partially filled outer shell. In metallic aluminum the three valence electrons per atom become conduction electrons. The number of conduction electrons is constant, depending on neither temperature nor impurities. Metals conduct electricity at all temperatures, but for most metals the conductivity is best at low temperatures. Divalent atoms, such as magnesium or calcium, donate both valence electrons to become conduction electrons, while monovalent atoms, such as lithium or gold, donate one. As will be recalled, the number of conduction electrons alone does not determine conductivity; it depends on electron mobility as well. Silver, with only one conduction electron per atom, is a better conductor than aluminum with three, for the higher mobility of silver compensates for its fewer electrons.

In metals such as sodium and aluminum, the atoms donate all their valence electrons to the conduction band. The resulting ions are small, occupying only 10–15 percent of the volume of the crystal. The conduction electrons are free to roam through the remaining space. A simple model, which often describes well the properties of the conduction electrons, treats them as interacting neither with the ions nor with each other. The electrons are approximated as free particles wandering easily through the crystal. This concept was first proposed by the German scientist Arnold Johannes Wilhelm Sommerfeld. It works quite well for those metals, known as simple metals, whose conduction electrons are donated from *sp*-shells—for example, aluminum, magnesium, calcium, zinc, and lead. They are called simple because they are aptly described by the simple theory of Sommerfeld.

The transition metals are found in three rows of the periodic table: the first row consists of scandium through nickel, the second row is yttrium through palladium, and the third row is lanthanum plus hafnium through platinum. Within these rows, as the atomic number increases, the electrons fill *d*-states in the outer shell of the atom. In crystal form the transition metal atoms are metals with interesting properties. The *d*-electrons are more tightly bound to the ion centre than are *sp*-electrons. While the *sp*-valence electrons become conduction electrons that move freely through the crystal, the *d*-electrons tend to stay localized near the ion. Neighbouring ions may covalently bond *d*-electrons. In most cases, these *d*-states are only partially filled. Electrons in these *d*-states can conduct as well as those in the *sp*-states, but the electron motion in the *d*-states is not well approximated by the Sommerfeld model of free particles. Instead, the electrons move from ion to ion through the shared covalent bonds of the *d*-electrons. These metals have some conduction electrons donated from *sp*-states and others from *d*-states; therefore, some electrons move freely according to the Sommerfeld model, while others move through the bonds. Each electron switches back and forth between these two modes of conduction, resulting in electron motion that is quite complicated.

An applied voltage causes the electrons of metals to accelerate and contribute to the electric current. The electrons scatter occasionally from imperfections in the crystal, and the rate of scattering determines the mobility. The electrons do not scatter from the ions in the crystal that are located at the expected site in the crystal lattice. The electrons move to accommodate the host ions rather than scatter from them. If an ion is missing, misplaced, or of a different species, however, the electron will scatter from this defect. Ions vibrate around their lattice site, with the amplitude vibration increasing with temperature. The vibration may cause the ion to be displaced from its crystal site, providing a defect from which an electron will scatter. The resistivity of metals increases at high temperature, owing to the increase in vibrations of the ions in the crystal and the resulting increase in scattering.

Conducting Properties of Semiconductors

Semiconductors have conducting properties intermediate to those of insulators and metals. In some cases the semiconductors are insulators, while in others they are metals. Semiconductors share with insulators the property that they have no conduction electrons in a perfect crystal without thermal fluctuations. Conduction electrons are provided by electrons from impurities or by thermal fluctuation of electrons from atomic shells. The important difference between insulators and semiconductors is in the nature of the traps. A trap is a local electron energy state at a defect. Although the traps in insulators bind conduction electrons tightly, those in semiconductors only weakly bind the electrons. A trapped conduction electron in a semiconductor can be kicked back to the conduction band by thermal fluctuations. At room temperature, the majority of extra electrons are found in the conduction band rather than in traps. The inability of traps to keep electrons is the main difference between semiconductors and insulators. A semiconductor at room temperature has a sufficient number of conduction electrons to provide good electrical conductivity. Since the mobility of electrons in many semiconductors is exceptionally high, even a small number of conduction electrons is generally sufficient to allow high conductivity.

Phosphorus has five valence electrons, while silicon has four. When a phosphorus atom substitutes for an atom in a silicon crystal lattice, four of its five valence electrons enter covalent bonds. The fifth one is extra, sitting in a shallow trap around the phosphorus site. It is easily excited, however, to the conduction band by thermal fluctuations. At room temperature, there is nearly one conduction electron in silicon for each phosphorus impurity. By controlling the number of impurities, it is possible to control the conductivity of silicon. Other substitutional atoms such as arsenic and antimony also serve as electron donors to the conduction band of silicon.

If a sufficient number of conduction electrons are added to a semiconductor through the introduction of impurities, the electrical properties become metallic. There is a critical concentration of impurities N_c, which depends on the type of impurity. For impurity concentrations less than the critical amount N_c, the conduction electrons become bound in traps at extremely low temperatures, and the semiconductor becomes an insulator. For a concentration of impurities higher than N_c, the conduction electrons are not bound in traps at low temperatures, and the semiconductor exhibits metallic conduction. For phosphorus impurities in silicon, $N_c = 2 \times 10^{18}$ impurities per cubic centimetre. Although this number seems large, it represents about one phosphorus atom for each 100,000 silicon atoms. On a percentage basis, a small number of phosphorus atoms will change silicon from an insulator to a metallic conductor. Other semiconductors have similar properties. In gallium arsenide the critical concentration of impurities for metallic conduction is 100 times smaller than in silicon.

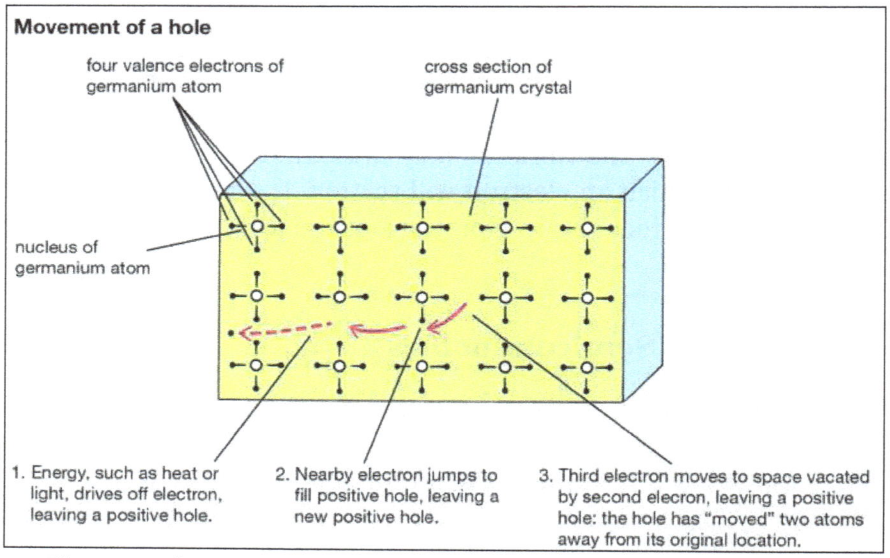

Electron Hole: Movement Movement of an electron hole in a crystal lattice.

Gallium atoms, like those of phosphorus, can be used as substitutional impurities in silicon. Each atom contributes three electrons to covalent bonds. Since four electrons are needed to complete a tetrahedral arrangement, there is one electron absent per gallium atom from a full set of covalent bonds. The missing electron is called a hole. Holes can move around the crystal in a process similar to the motion of ion vacancies, except in this case there is an electron vacancy. An electron from a nearby covalent bond can jump over and fill the empty electron state, thereby moving the hole to the neighbouring bond. The hole contributes a positive charge, since it is the absence of an electron. The mobility of holes in response to an external voltage is almost as high as the mobility of conduction electrons. A semiconductor may have a high density of impurities that cause holes, and a high electrical conductivity is created by their motion. A *p*-type semiconductor is one with a preponderance of holes; an *n*-type semiconductor has a preponderance of conduction electrons. The symbols *p* and *n* come from the sign of the charge of the particles: *p*ositive for holes and *n*egative for electrons.

Thermal fluctuations can excite an electron out of a covalent bond, making it a conduction electron. The bond is left with a missing electron, which constitutes a hole. Thermal fluctuations thus make electron-hole pairs. Usually the electron and hole separate in space, and each wanders away. The Swiss-American scientist Gregory Hugh Wannier first suggested that the electron and hole could bind together weakly. This bound state, called a Wannier exciton, does exist; the hole has a positive charge, the electron has a negative charge, and the opposites attract. The exciton is observed easily in experiments with electromagnetic radiation. It lives for only a short time—between a nanosecond and a microsecond—depending on the semiconductor. The short lifetime is due to the preference for the electron to reenter a covalent bond state, thereby eliminating both the hole and the conduction electron. This recombination of electron and hole is easily accomplished from the exciton state, since the two particles are spatially nearby. If the electron and hole escape the exciton state by thermal fluctuation, they travel away from each other. Recombination is then less probable, since it occurs only when the wandering particles pass close to one another again. Recombination also can occur at defect sites. First, one particle becomes bound to the defect, followed by the second particle. The electron and hole are again close to one another, and the electron can reoccupy the covalent bond.

As in metals, the mobility of electrons in semiconductors is limited by electron scattering. For crystals with few defects, the mobility is limited by defect scattering at the lowest temperatures and by ion vibrations at moderate and high temperatures. Since semiconductors with few defects have a small number of conduction electrons, the resistivity is high. The number of conduction electrons is increased in semiconductors by adding impurities. Unfortunately, this also increases the scattering from impurities, which reduces the mobility. Figure shows the resistivity of silicon at room temperature (T = 300 K) as a function of the concentration of impurities. The two curves represent conduction by electrons and by holes. Each grid mark on the graph is a factor of 10. The resistivity varies by a factor of one million from the lowest to the highest concentration of impurities.

Semiconductors with few impurities are good photoconductors. Photoconductivity is the phenomenon in which the electrical conductivity of a solid is increased by exposing it to light. Light is electromagnetic radiation within a specific narrow band of frequencies. The quanta of light are absorbed by the semiconductor, creating electron-hole pairs that provide the electrical conduction. More intense light produces more electron-hole pairs and gives rise to better conductivity. Each semiconductor absorbs light over a specific frequency range, so different semiconductors are used as photoconductors for different ranges of frequency.

Zinc oxide (ZnO) is an interesting material with respect to conductivity. It crystallizes in the wurtzite structure, and its bonding is a mix of ionic and covalent. High-purity single crystals are insulators. Zinc oxide is the most piezoelectric of all materials and is widely used as a transducer in electronic devices. (Piezoelectricity is the property of a crystal to become polarized when subjected to pressure.) Zinc oxide is a good semiconductor when aluminum impurities are included in the crystal. Polycrystalline ceramics of semiconducting zinc oxide conduct well and obey Ohm's law. The addition of small amounts of other oxides, such as those of barium and chromium, causes zinc oxide ceramics to have very nonohmic electrical properties; the electrical current in such ceramics is the most nonlinear of any known material. The current I becomes proportional to a power of the voltage V^n, where the exponent n has values of more than 100 in certain ranges of voltage. This material is called a varistor, which is a contraction of the words variable and resistor. Zinc oxide varistors are widely used as circuit elements to protect against voltage surges. Figure above shows a graph of current versus voltage for a zinc oxide varistor used in household electronics. There is little current until a critical voltage of about 330 volts is reached, at which point the current rises steeply in a nonlinear fashion. Another interesting application of zinc oxide was its former use as a white pigment in paint. It has been replaced by titanium dioxide (TiO_2), however, which is whiter.

Magnetism

Explanation of Magnetism

Electrons are perpetually rotating, and, since the electron has a charge, its spin produces a small magnetic moment. Magnetic moments are small magnets with north and south poles. The direction of the moment is from the south to the north pole. In nonmagnetic materials the electron moments cancel, since there is random ordering to the direction of the electron spins. Whenever two electrons have their moments aligned in opposite directions, their effects tend to cancel. Magnets are formed when a large number of the electrons align their individual moments in the same direction. Only a small percentage of crystals are magnetic. The forces that tend to align the electron spins are subtle. There are three separate parts to the explanation of magnetism. They are as follows:

1. Most magnets are composed of atoms whose valence electrons are in *d*- or *f*-shells. Atomic shell notation refers to angular momentum, where *s* has zero unit, *p* has one, *d* has two, and *f* has three. Electrons in *d*-shells tend to be bound to the ion, and those in *f*-shells are bound even more tightly.

2. Each electron orbital can be occupied by two electrons—one with spin up and one with spin down. The *d*-shell has five orbital states and 10 electrons when filled; the *f*-shell has seven orbital states and 14 electrons when filled. Electrons are added one at a time to the *d*-states according to the empirical rule that the electrons arrange themselves in the state with the maximum spin and the maximum magnetic moment. If the first electron has spin up, the next four will also have spin up. A maximum of five electrons with spin up are allowed in the *d*-shell, so the sixth must have spin down. Similarly, the *f*-shell accepts seven electrons with the same direction of spin before taking electrons with the opposite spin orientation. The order in which electrons fill atomic shells is described by Hund's rules, of which the first is maximizing the total spin. Atoms with electrons in partially filled *d*- and *f*-shells usually have a nonzero total spin and thus a net magnetic moment. These magnetic ions are the building blocks for magnetic crystals.

Hund's first rule is due to a phenomenon called electron exchange. As discussed above, a fundamental rule of quantum mechanics, the Pauli exclusion principle, states that no two electrons with the same direction of spin can occupy the same point in space at the same time. Electrons have charge and repel one another. If two electrons come close together, a large amount of repulsive energy is produced. Physical systems prefer the state of lowest energy, and so electrons avoid such close approach. When their spins are parallel, electrons avoid each other because of the Pauli principle. Electrons in the same shell thus prefer to have their spins parallel, since this configuration keeps the electrons apart and thereby reduces the amount of repulsive energy. The concept of electron exchange is the basis of magnetism. It explains why ions such as iron have large magnetic moments. In divalent iron (Fe^{2+}) the six *d*-electrons are arranged to achieve maximum electron spin and magnetic moment.

3. Individual ions with fixed magnetic moments may cooperatively align their moments, resulting in the presence of magnetic properties of the crystal as a whole. Ferromagnet crystals have the magnetic moments from all their constituent ions aligned in the same direction; the magnetic moment of the crystal is the summation of the individual moments of the ions. There must be a magnetic force between the different ions that causes them to cooperatively align their moments. This force is also due to electron exchange. The *d*-orbitals from neighbouring ions overlap weakly into covalent bonds. The *d*-electrons on the separate ions are shared with the neighbour through covalent bonding. The electron exchange will tend to align the spins on the two neighbours. Aligning all pairs of neighbours aligns all ions. The exchange force between neighbours is much weaker than the force within the atomic shell of one ion. Although weak, the force is sufficient to cause ferromagnetism.

Ferromagnetic Materials

Iron is a typical ferromagnet. Not all bars of iron are magnets; the existence of magnetism is determined by the nature of the domains within the bar. A domain is a region of a crystal in which all the ions are ferromagnetically aligned in the same direction. A bar may be composed of many domains, each having a different magnetic orientation. Such a bar would not appear to be magnetic. Each piece of the bar is magnetic, but the domains have moments that point in different directions, so the bar has no net moment. If the bar of iron is placed in a strong magnetic field, however, the bar becomes magnetic. The field causes the bar to become a single domain with all moments aligned along the external field. The domains do not rotate their moments; instead, the walls between domains move. The domain with a moment along the field grows, while the others become smaller. If removed from the magnetic field, the iron bar will remain magnetized for a considerable time period. Nearly all bars of iron are polycrystalline: they have many small grains of single crystals, which are packed together with random orientation. A grain could be a single domain, a domain could include many grains, or a large grain could have several domains.

Ferromagnetic materials change their magnetic ordering at a characteristic temperature T_c called the Curie temperature. The Curie temperatures for three common ferromagnetics—iron, cobalt, and nickel—are 1,043 K; 1,394 K; and 631 K, respectively. For temperatures below T_c the magnetic moments of the ions are aligned and the crystal is magnetic. For temperatures above T_c the crystal is not ferromagnetic, since the individual atomic moments are no longer aligned. Above T_c the moments have short-range order but not long-range order. Short-range order means there is local ordering. If a moment points in one direction, its neighbours have a tendency to point in the same direction. This tendency is maintained over several lattice sites but is not maintained for long distances. Long-range order is the tendency for moments to align for large distances. For temperatures a few degrees below T_c the moments have strong short-range order but only a small amount of long-range order, so the bar is not very magnetic. The tendency for long-range order increases at lower temperature. The Curie temperature is the point where long-range order begins as the temperature is lowered.

If an iron bar is heated to a temperature above T_c, the bar is no longer magnetic. If the bar is then cooled to a temperature below T_c, the grains become magnetic, but they orient their moments in random directions, so the bar as a whole is not magnetic. A bar can be demagnetized by heating the bar and then cooling it. By inserting it in a large magnetic field, the bar can be remagnetized.

Ferromagnetism is found in many insulators as well as metals. Chromium bromide ($CrBr_3$) is an insulator since chromium is trivalent and a bromine atom needs one electron to complete its outer shell. The trivalent chromium atoms each have a moment, and these align ferromagnetically below the Curie temperature of 37 K. Gadolinium chloride ($GdCl_3$; T_c = 2.2 K) and europium oxide (EuO; T_c = 77 K) are two other examples among many.

Antiferromagnetic Materials

Many crystals have magnetic ions that are ordered in arrangements other than ferromagnetic. In antiferromagnetic ordering, the moments pointing in one direction are balanced by others pointing in the opposite direction, with the result that the substance has no net magnetization. The exchange interaction between ions in this case has the opposite sign and favours the alternate arrangements of spins. The sign of the exchange interaction between ions depends on the length of the covalent bond and the bonding angles; it may have either orientation. The characteristic temperature associated with antiferromagnetism is called the Néel temperature T_N. Below T_N the ions are antiferromagnetically ordered, while above this temperature there is no long-range antiparallel order. Some examples of antiferromagnetic crystals are manganese oxide (MnO; T_N = 116 K), manganese sulfide (MnS; T_N = 160 K), and iron oxide (FeO; T_N = 198 K). Manganese oxide is an insulator since manganese atoms are divalent and oxygen atoms accept two electrons. The manganese ion has a fixed magnetic moment. The crystal structure of manganese oxide is the same as that of sodium chloride shown in figure. Below the Néel temperature the atomic unit cell doubles in size to include two atoms of each type of ion. This is necessary because below T_N neighbouring manganese atoms have moments in the opposite direction and are no longer equivalent; the unit cell must therefore include one moment in each of the two directions. Fluorides such as manganese fluoride (MnF_2), iron (II) fluoride (FeF_2), cobalt fluoride (CoF_2), and nickel fluoride (NiF_2) are other crystals that exhibit antiferromagnetic ordering of the transition metal ions.

Ferrimagnetic Materials

Ferrimagnetism is another type of magnetic ordering. In ferrimagnets the moments are in an antiparallel alignment, but they do not cancel. The best example of a ferrimagnetic mineral is magnetite (Fe_3O_4). Two iron ions are trivalent, while one is divalent. The two trivalent ions align with opposite moments and cancel one another, so the net moment arises from the divalent iron ion. The historic lodestone that was the first magnetic material discovered was a form of magnetite.

Another class of ferrimagnets has the garnet structure; $Y_3Fe_5O_{12}$ is one such crystal. Only the iron ions are magnetic. Three point in one direction and two in the other, so there is a net magnetic moment of one iron ion in the unit cell. However, if a rare-earth ion such as gadolinium (Gd) is substituted for yttrium (Y), then the rare-earth ion also contributes to the ferrimagnetism.

Ferrites are oxides of iron with the formula MFe_2O_4, where M is a divalent ion such as nickel, zinc, cadmium, manganese, or magnesium. The ferric iron ion is trivalent, and the oxygen ion accepts two electrons. Actually M can also be divalent iron, forming magnetite (Fe_3O_4). The crystal structure is called spinel, which is the mineral name for $MgAl_2O_4$. Ferrites are electrical insulators with magnetic ordering. Their insulating quality makes them useful as magnetic cores. When metallic ferromagnetic materials are exposed to alternating magnetic fields, significant heating losses occur from eddy currents. Ferrite magnets greatly reduce such heat losses because of their high resistivity. They also absorb electromagnetic radiation of very long wavelengths. This property is unusual, since metals reflect such radiation while insulators transmit it. In small crystallites of ferrites, the electromagnetic radiation causes the magnetic moment to rotate at the frequency of light. The absorption frequency of the light depends on the detailed shape of the crystal grain. A polycrystalline material, with a range of grain sizes and shapes, absorbs over a broad range of radar frequencies. Ferrite crystals are a major ingredient in snooper paint, which makes stealth airplanes undetectable by radar. Regular airplanes reflect radar, and the reflected signals can be used to locate the aircraft. Stealth planes absorb the radar signals and so cannot be located in this way.

The Kondo Effect

Magnetic ions have interesting properties when they are found as impurities in nonmagnetic crystals. They usually retain their magnetic moment, so small magnets are distributed randomly throughout the crystal. If the host crystal is a metal, the magnetic impurities make an interesting contribution to the electrical resistivity. The conduction electrons scatter from the magnetic impurity. Since the conduction electron and the impurity both have spin, they can mutually flip spins while scattering. The spin-flip scattering is strong at low temperatures and actually increases slightly as temperature decreases. This phenomenon is called the Kondo effect after the Japanese theoretical physicist Jun Kondo, who first explained the increase in resistivity resulting from magnetic impurities. There is a characteristic temperature, called the Kondo temperature, which depends on the impurity and on the metallic host. The resistivity increases at low temperature, starting near the Kondo temperature. A typical example of a Kondo system is iron impurities in copper; the system's Kondo temperature is 24 K. The solid line in the resistivity in copper at low temperature when there are 110 iron impurities per 1,000,000 copper atoms. The dashed line a is the resistivity in the absence of impurities. It increases at higher temperature because the electron scatters from ion vibrations. The dashed line b is the resistivity from spin-flip scattering. Nearly all transition metal atoms are found as magnetic impurities in copper or gold. Each system has a different Kondo temperature, which varies from 1,000 K to a fraction of 1 K. The spin-flip part of the electrical resistivity is unique in that it is large at low temperatures and decreases at high temperatures; most contributions to the resistivity increase at high temperatures.

Crystal Habit

A mineralogist often describes a mineral in terms associated with the apparent shape and size of its crystals. For example, a branching structure is described as *dendritic*; a star-like, radiating form is called *stellate*; a structure with needle-shaped crystals is called *acicular*. Such a description is known as the crystal habit of the mineral. A list of crystal habits is given below.

Pyrite sun (or dollar) in laminated shale matrix. Between tightly spaced layers of shale, the aggregate was forced to grow in a laterally compressed, radiating manner. Under normal conditions, pyrite would form cubes or pyritohedrons.

The various terms used for crystal habits are useful in communicating the appearance of mineral specimens. Recognizing numerous habits helps a mineralogist identify a large number of minerals. Some habits are distinctive of certain minerals, but most minerals exhibit differing habits that are influenced by certain factors. Crystal habit may mislead the inexperienced person, as a mineral's crystal system can be hidden or disguised.

Factors influencing a crystal's habit include: a combination of two or more forms; trace impurities present during growth; and growth conditions, such as heat, pressure, and space available for growth. Minerals belonging to the same crystal system do not necessarily exhibit the same habit.

Some habits of a mineral are unique to its variety and locality. For example, while most sapphires form elongate, barrel-shaped crystals, those found in Montana form stout, *tabular* crystals. Ordinarily, the latter habit is seen only in ruby. Sapphire and ruby are both varieties of the same mineral, corundum.

Sometimes, one mineral may replace another, while preserving the original mineral's habit. This process is called *pseudomorphous replacement*. A classic example is tiger's eye quartz, in which silica replaces crocidolite asbestos. Quartz typically forms *euhedral* (well-formed), *prismatic* (elongate, prism-like) crystals, but in the case of tiger's eye, the original, *fibrous* habit of crocidolite is preserved.

List of Crystal Habits

Habit:	Description:	Example:
Acicular	Needle-like, slender and/or tapered	Rutile in quartz
Amygdaloidal	Almond-shaped	Heulandite
Anhedral	Poorly formed, external crystal faces not developed	Olivine

Bladed	Blade-like, slender and flattened	Kyanite
Botryoidal or globular	Grape-like, hemispherical masses	Smithsonite
Columnar	Similar to fibrous: Long, slender prisms often with parallel growth	Calcite
Coxcomb	Aggregated flaky or tabular crystals closely spaced.	Barite
Dendritic or arborescent	Tree-like, branching in one or more directions from central point	Magnesite in opal
Dodecahedral	Dodecahedron, 12-sided	Garnet
Drusy or encrustation	Aggregate of minute crystals coating a surface	Uvarovite
Enantiomorphic	Mirror-image habit and optical characteristics; right- and left-handed crystals	Quartz
Equant, stout, stubby or blocky	Squashed, pinnacoids dominant over prisms	Zircon
Euhedral	Well-formed, external crystal faces developed	Spinel
Fibrous or columnar	Extremely slender prisms	Tremolite
Filiform or capillary	Hair-like or thread-like, extremely fine	Natrolite
Foliated or micaceous	Layered structure, parting into thin sheets	Mica
Granular	Aggregates of anhedral crystals in matrix	Scheelite
Hemimorphic	Doubly terminated crystal with two differently shaped ends.	Hemimorphite
Mamillary	Breast-like: intersecting large rounded contours	Malachite
Massive or compact	Shapeless, no distinctive external crystal shape	Serpentine
Nodular or tuberose	Deposit of roughly spherical form with irregular protuberances	Geodes
Octahedral	Octahedron, eight-sided (two pyramids base to base)	Diamond
Plumose	Fine, feather-like scales	Mottramite
Prismatic	Elongate, prism-like: all crystal faces parallel to c-axis	Tourmaline
Pseudo-hexagonal	Ostensibly hexagonal due to cyclic twinning	Aragonite
Pseudomorphous	Occurring in the shape of another mineral through pseudomorphous replacement	Tiger's eye
Radiating or divergent	Radiating outward from a central point	Pyrite suns
Reniform or colloform	Similar to mamillary: intersecting kidney-shaped masses	Hematite
Reticulated	Acicular crystals forming net-like intergrowths	Cerussite
Rosette	Platy, radiating rose-like aggregate	Gypsum
Sphenoid	Wedge-shaped	Sphene
Stalactitic	Forming as stalactites or stalagmites; cylindrical or cone-shaped	Rhodochrosite
Stellate	Star-like, radiating	Pyrophyllite
Striated/striations	Surface growth lines parallel or perpendicular to c-axis	Chrysoberyl
Subhedral	External crystal faces only partially developed	
Tabular or lamellar	Flat, tablet-shaped, prominent pinnacoid	Ruby
Wheat sheaf	Aggregates resembling hand-reaped wheat sheaves	Zeolites

Uses of Crystals

Historically, gemstones, which are natural crystals, have been sought after for their aesthetic appeal. In addition, they have been said to possess healing properties. Crystals (both natural and synthetic) also have a variety of practical applications, some of which are noted below.

- Solid-state laser materials are often made by doping a crystalline solid with appropriate ions. For example, the first working laser was made from a synthetic ruby crystal (chromium-doped corundum). Also, titanium-doped sapphire (corundum) produces a highly tunable infrared laser.
- Mica crystals, which are excellent as electrical insulators, are used in the manufacture of capacitors and insulation for high-voltage electrical equipment.
- Based on their extreme hardness, diamonds are ideal for cutting, grinding, and engraving tools. They can be used to cut, polish, or wear away practically any material, including other diamonds.
- Quartz crystals, which have piezoelectric properties, are commonly used to make "oscillators" that keep track of time in wristwatches, provide a stable clock signal for digital integrated circuits, and stabilize radio transmitter frequencies.

Crystalline Materials

Crystalline structures occur in all classes of materials, with all types of chemical bonds. Almost all metals exist in a polycrystalline state. Amorphous or single-crystal metals may be produced synthetically, often with great difficulty. Ionically bonded crystals are often formed from salts, when the salt is solidified from a molten fluid or when it is crystallized out of a solution. Covalently bonded crystals are also common, notable examples being diamond, silica, and graphite. Weak interactions, known as Van der Waals forces, can also play a role in a crystal structure; for example, this type of bonding loosely holds together the hexagonal-patterned sheets in graphite. Polymers generally form crystalline regions, but the lengths of the molecules usually prevents complete crystallization.

Some crystalline materials may exhibit special electrical properties, such as the ferroelectric effect or the piezoelectric effect. Additionally, light passing through a crystal is often bent in different directions, producing an array of colors. The study of these effects is called *crystal optics*.

Most crystalline materials have a variety of crystallographic defects. The types and structures of these defects can have a profound effect on the properties of the materials.

2

Understanding Crystallography

Crystallography is the branch of chemistry which is concerned with the arrangement of atoms in a crystal. Symmetry in crystallography is useful for the characterization of different crystals and its physical properties such as thermal conductivity and optical activity. This chapter has been carefully written to provide an easy understanding of the varied facets of crystallography.

Crystallography is branch of science that deals with discerning the arrangement and bonding of atoms in crystalline solids and with the geometric structure of crystal lattices. Classically, the optical properties of crystals were of value in mineralogy and chemistry for the identification of substances. Modern crystallography is largely based on the analysis of the diffraction of X-rays by crystals acting as optical gratings. Using X-ray crystallography, chemists are able to determine the internal structures and bonding arrangements of minerals and molecules, including the structures of large complex molecules, such as proteins and DNA.

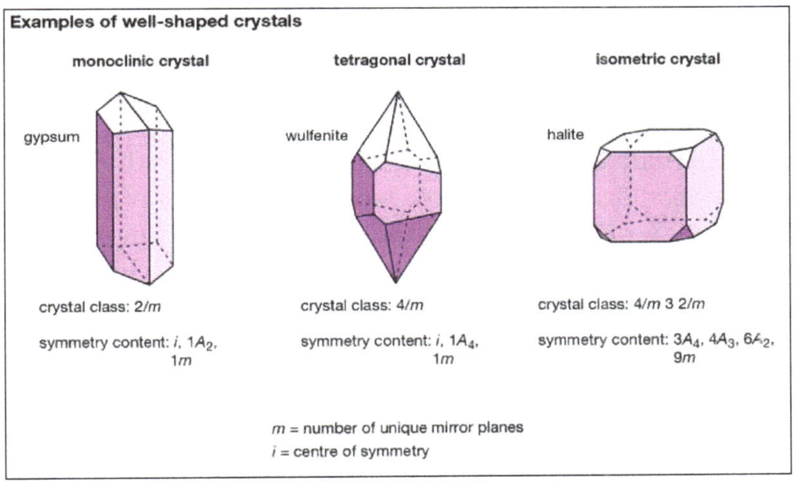

Examples of well-shaped crystals.

Crystallographic methods now depend on the analysis of the diffraction patterns that emerge from a sample that is targeted by a beam of some type. The beam is not always electromagnetic radiation, although X-rays are the most common choice. For some purposes, electrons or neutrons are used, which is possible due to the wave properties of the particles. Crystallographers often explicitly state the type of illumination used when referring to a method, as with the terms X-ray diffraction, neutron diffraction, and electron diffraction.

Theory

An image of a small object is usually generated by using a lens to focus the illuminating radiation, as is done with the rays of the visible spectrum in light microscopy. However, the wavelength of visible light (about 4000 to 7000 Angstroms) is three orders of magnitude longer then the length of typical atomic bonds and atoms themselves (about 1 to 2 Angstroms). Therefore, obtaining information about the spatial arrangement of atoms requires the use of radiation with shorter wavelengths, such as X-rays. Employing shorter wavelengths implied abandoning microscopy and true imaging, however, because there exists no material from which a lens capable of focusing this type of radiation can be created. (That said, scientists have had some success focusing X-rays with microscopic Fresnel zone plates made from gold, and by critical-angle reflection inside long tapered capillaries). Diffracted x-ray beams cannot be focussed to produce images, so the sample structure must be reconstructed from the diffraction pattern. Sharp features in the diffraction pattern arise from periodic, repeating structure in the sample, which are often very strong due to coherent reflection of many photons from many regularly spaced instances of similar structure, while non-periodic components of the structure result in diffuse (and usually weak) diffraction features. Because of their highly ordered and repetitive structure, crystals give diffraction patterns of sharp Bragg reflection spots, and are ideal for analyzing the structure of solids.

Technique

Some materials studied using crystallography, proteins for example, do not occur naturally as crystals. Typically, such molecules are placed in solution and allowed to crystallize over days, weeks, or months through vapor diffusion. A drop of solution containing the molecule, buffer, and precipitants is sealed in a container with a reservoir containing a hygroscopic solution. Water in the drop diffuses to the reservoir, slowly increasing the concentration and allowing a crystal to form. If the concentration were to rise more quickly, the molecule would simply precipitate out of solution, resulting in disorderly granules rather than an orderly and hence usable crystal.

Once a crystal is obtained, data can be collected using a beam of radiation. Although many universities that engage in crystallographic research have their own X-ray producing equipment, synchrotrons are often used as X-ray sources, because of the purer and more complete patterns such sources can generate. Synchrotron sources also have a much higher intensity of X-ray beams, so data collection takes a fraction of the time normally necessary at weaker sources.

Producing an image from a diffraction pattern requires sophisticated mathematics and often an iterative process of modeling and refinement. In this process, the mathematically predicted diffraction patterns of an hypothesized or "model" structure are compared to the actual pattern generated by the crystalline sample. Ideally, researchers make several initial guesses, which through refinement all converge on the same answer. Models are refined until their predicted patterns match to as great a degree as can be achieved without radical revision of the model. This is a painstaking process, made much easier today by computers.

The mathematical methods for the analysis of diffraction data only apply to *patterns,* which in turn result only when waves diffract from orderly arrays. Hence crystallography applies for the most part only to crystals, or to molecules which can be coaxed to crystallize for the sake of measurement. In spite of this, a certain amount of molecular information can be deduced from the patterns that are generated by fibers and powders, which while not as perfect as a solid crystal, may exhibit a degree of order. This level of order can be sufficient to deduce the structure of simple molecules, or to determine the coarse features of more complicated molecules (the double-helical structure of DNA, for example, was deduced from an X-ray diffraction pattern that had been generated by a fibrous sample).

Materials Engineering

Crystallography is a tool that is often employed by materials scientists. In single crystals, the effects of the crystalline arrangement of atoms is often easy to see macroscopically, because the natural shapes of crystals reflect the atomic structure. In addition, physical properties are often controlled by crystalline defects. The understanding of crystal structures is an important prerequisite for understanding crystallographic defects. Mostly, materials do not occur in a single crystalline, but poly-crystalline form, such that the powder diffraction method plays a most important role in structural determination.

A number of other physical properties are linked to crystallography. For example, the minerals in clay form small, flat, platelike structures. Clay can be easily deformed because the platelike particles can slip along each other in the plane of the plates, yet remain strongly connected in the direction perpendicular to the plates. Such mechanisms can be studied by crystallographic texture measurements.

In another example, iron transforms from a body-centered cubic (bcc) structure to a face-centered cubic (fcc) structure called austenite when it is heated. The fcc structure is a close-packed structure, and the bcc structure is not, which explains why the volume of the iron decreases when this transformation occurs.

Crystallography is useful in phase identification. When performing any process on a material, it may be desired to find out what compounds and what phases are present in the material. Each phase has a characteristic arrangement of atoms. Techniques like X-ray diffraction can be used to identify which patterns are present in the material, and thus which compounds are present (note: the determination of the "phases" within a material should not be confused with the more general problem of "phase determination," which refers to the phase of waves as they diffract from planes within a crystal, and which is a necessary step in the interpretation of complicated diffraction patterns).

Crystallography covers the enumeration of the symmetry patterns which can be formed by atoms in a crystal and for this reason has a relation to group theory and geometry.

Biology

X-ray crystallography is the primary method for determining the molecular conformations of biological macromolecules, particularly protein and nucleic acids such as DNA and RNA. In fact, the double-helical structure of DNA was deduced from crystallographic data. The first crystal structure of a macromolecule, myoglobin, was solved in 1958.

The Protein Data Bank (PDB) is a freely accessible repository for the structures of proteins and other biological macromolecules. Computer programs like RasMol or Pymol can be used to visualize biological molecular structures.

Electron crystallography has been used to determine some protein structures, most notably membrane proteins and viral capsids.

Notations used in Crystallography

- Coordinates in square brackets such as denote a direction vector (in real space).

- Coordinates in *angle brackets* or *chevrons* such as <100> denote a *family* of directions which are related by symmetry operations. In the cubic crystal system for example, <100> would mean,, or the negative of any of those directions.
- Miller indices in *parentheses* such as (100) denote a plane of the crystal structure, and regular repetitions of that plane with a particular spacing. In the cubic system, the normal to the (hkl) plane is the direction [hkl], but in lower-symmetry cases, the normal to (hkl) is not parallel to [hkl].
- Indices in *curly brackets* or *braces* such as {100} denote a family of planes and their normals which are equivalent due to symmetry operations, much the way angle brackets denote a family of directions.

Symmetry in Crystallography

In crystallography, symmetry is used to characterize crystals, identify repeating parts of molecules, and simplify both data collection and nearly all calculations. Also, the symmetry of physical properties of a crystal such as thermal conductivity and optical activity must include the symmetry of the crystal. Thus, a *thorough* knowledge of symmetry is essential to a crystallographer.

An object is described as *symmetric* with respect to a *transformation* if the object appears to be in a state that is identical to its initial state, after the transformation. In crystallography, most types of symmetry can be described in terms of an apparent *movement* of the object such as some type of rotation or translation. The apparent movement is called the symmetry *operation*. The locations where the symmetry operations occur such as a rotation axis, a mirror plane, an inversion center, or a translation vector are described as *symmetry elements*.

Two distinct methods of describing rotational symmetry operations exist. These two sets of descriptors are the *Hermann-Mauguin* nomenclature and the *Schönflies* nomenclature. The Carl Hermann- Charles Mauguin system is typically used to describe crystals and crystallographic symmetry. The Arthur Schönflies convention is primarily used to describe symmetry in discrete molecules, in optical spectroscopy, and in quantum mechanics. Hermann-Mauguin notation will be listed first followed by the corresponding Schönflies notation in parentheses.

Symmetry in Crystals

Our discussion of symmetry in crystallography should begin with a description of crystals. Crystals are defined as solids that have an atomic structure with long-range, 3-dimensional order. Unfortunately, this long-range order cannot be absolutely confirmed by any other method than some diffraction technique. However, there are several observations that can be made that will strongly suggest that a sample is *crystalline* before a diffraction experiment is undertaken.

Typically, crystals have flat faces and sharp edges. Also, many crystals will have one or more directions that can be cleaved cleanly. Samples with a naturally round shape, or samples that have a concoidal fracture pattern are nearly always described as a glass having no significant, long-range,

3-D order. Similarly, materials that can be gently poked with a probe, and retain the deformed shape are gels or plastic materials and hence have not long-range, 3-D order.

When you look at several crystals from one material, you will soon notice that, although the crystals may have different sizes, all crystals have the same shape or *habit*. In particular, the angles between certain pairs of faces of the different crystals will be the same. This observation was first made by Nicholas Steno in 1669. This observation became known as the law of constancy of interfacial angles.

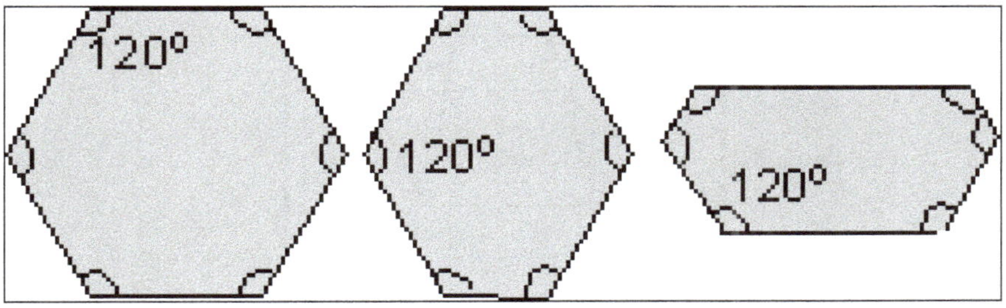

Models of crystals showing constant angles.

Steno and others in the 17th century were interested in the specific make up of crystals that would allow them to maintain the same angles between pairs of faces. These scientists believed that crystals must be made of some regularly-repeating components. Through these early studies, René-Just Haüy was able to postulate that if crystals of calcite and cubic garnets were built from many small blocks, then these blocks could easily be used to describe the faces of these crystals in terms of rational indices.[5] This law of rational indices forms the basis of optical crystallography.

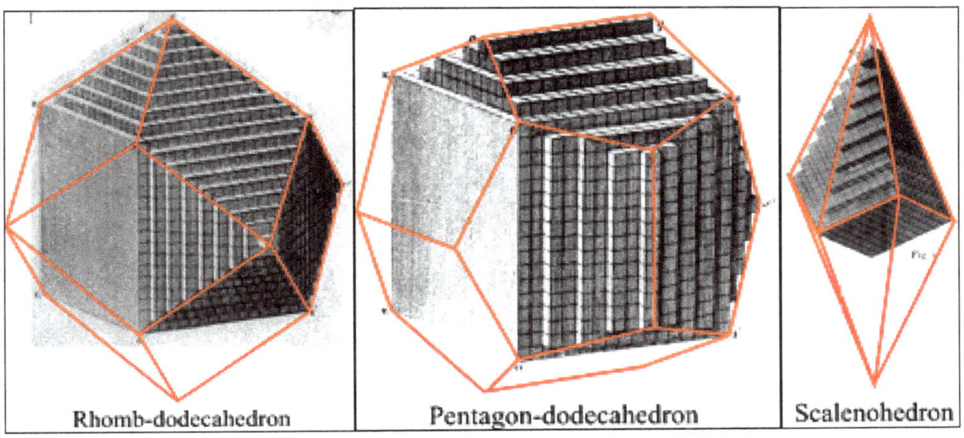

The crystal form have been redrawn in red.

Unit Cells

These regularly-repeating blocks are now known as *unit cells*. The dimensions of a unit cell are described by the lengths of the three axes, *a*, *b*, and *c*, and the three interaxial angles, α, β, and γ. In most published papers the axial lengths are expressed in terms of Å (Ångströms), and the interaxial angles are expressed in terms of ° (degrees).

There are many choices of repeating blocks in any given lattice. The main principles defining the lattice is that each lattice point must be in an identical environment as any other lattice point, and that the individual blocks in the lattice must have the smallest volume possible. Often there are many ways to select the vectors between lattice points and even the locations of the lattice points themselves. These unique lattice vectors are called *basis* vectors or *basis* set.

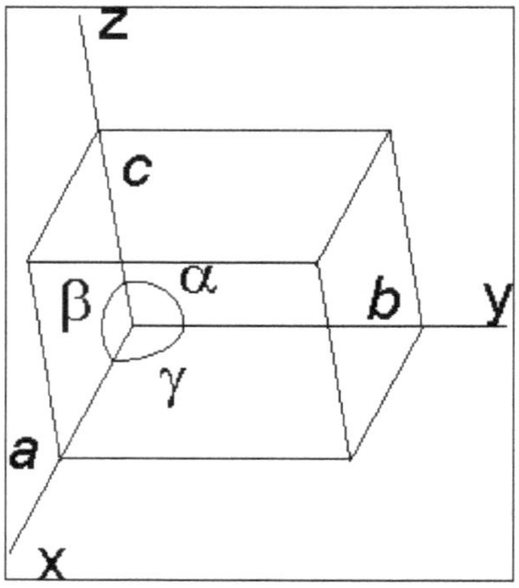

Unit cell showing cell parameters.

When researchers discuss a particular material, they need to work from one standard or conventional description of the unit cell for that material. Thus crystallographers have chosen the following criteria for selecting unit cells. By convention the unit cell edges are chosen to be *right-handed* ($a \times b$ is the direction of c), to have the *highest symmetry*, and to have the *smallest cell volume*. If other symmetry considerations do not override, then the cell is chosen so that $a \leq b \leq c$, and α, β, and γ all < 90° or all ≥ 90°. This type of cell is called the *reduced* cell. There are several other rules for obtaining the conventional *reduced cell* for a given material.

Crystalline materials are separated into 7 crystal different systems. These crystal systems are most easily identified by the constraints on the cell parameters. Note, however, that the cell parameter constraints are only necessary conditions. Thus a particular sample could have cell parameters that appear to fall into one category within experimental error, but are actually of lower symmetry.

The 7 crystal systems are listed. In the lowest symmetry system, triclinic, there are no restrictions on the values of the cell parameters. In the other crystal systems, symmetry reduces the number of unique lattice parameters as shown. Certain conventions have been followed in tabulating the parameters. In the monoclinic system, one of the axes is unique in the sense that it is perpendicular to the other two axes. This axis is selected by convention as either the b or c axis so that either β or γ are ≥ 90°, respectively. Note that c-unique monoclinic cells are common in French literature and b-unique cells are common in most other languages. In the tetragonal, trigonal, and hexagonal systems, one axis contains higher symmetry. By convention this axis is selected as the c axis.

Table: Crystal Systems.

Crystal System	#	Cell Parameters	Symmetry
Triclinic	6	$a \neq b \neq c; \alpha \neq \beta \neq \gamma$	$\bar{1}$
Monoclinic	4	$a \neq b \neq c;\ \alpha = \gamma = 90°,\ \beta \geq 90°$	2/m

Orthorhombic	3	$a \neq b \neq c;\ \alpha = \beta = \gamma = 90°$	mmm
Tetragonal	2	$a = b \neq c;\ \alpha = \beta = \gamma = 90°$	4/m, 4/mmm
Trigonal			
Hexagonal	2	$a = b \neq c; \alpha = \beta = 90°,\ \gamma = 120°$	$\bar{3}$, $\bar{3}$m
Rhombohedral	2	$a = b = c; \alpha = \beta = \gamma \neq 90°$	$\bar{3}$, $\bar{3}$m
Hexagonal	2	$a = b \neq c;\ \alpha = \beta = 90°, \gamma = 120°$	6/m, 6/mmm
Cubic	1	$a = b = c;\ \alpha = \beta = \gamma = 90°$	2/m$\bar{3}$, m$\bar{3}$m

The seven crystal systems each describe separate ways that simple 3-dimensional lattices may be constructed. As with all lattice systems, crystalline lattices are considered to have lattice points on the corners of the unit cell. Lattice points are selected so that the local environment around any particular lattice point is identical to the environment around any other lattice point.

Some trigonal lattices can be expressed on the basis of either a hexagonal or rhombohedral lattice. These lattices are shown in the drawings below. Note that rhombohedral lattice vectors can be expressed in an obverse orientation a), or in a reverse orientation b) below.

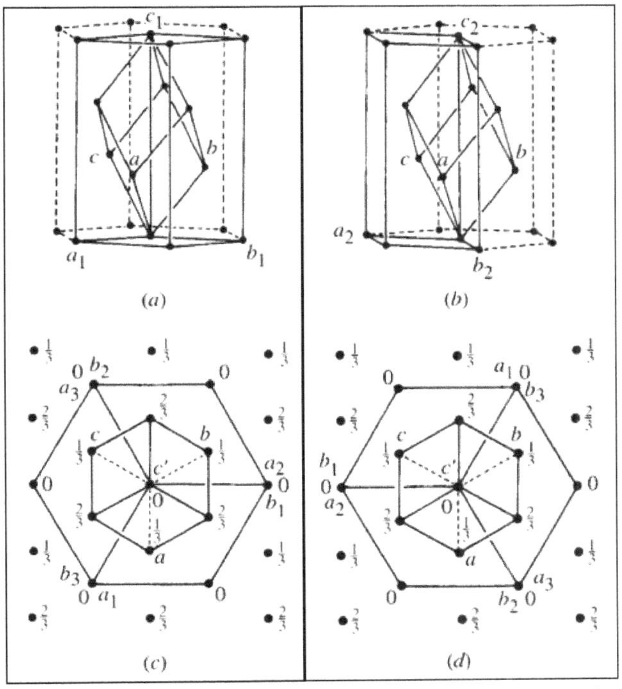

Rhombohedral lattices in the obverse and reverse orientations.

Bravais Lattices

It is sometimes possible to generate a lattice with higher symmetry if the lattice vectors are chosen so that one or more lattice points are also on the center of a face of the lattice or inside of the unit cell. These lattices with additional lattice points are described as centered lattices. Lattices with lattice points only on the corners are called primitive and are designated with the symbol P. Note that reduced cells described above are always primitive. In 1849, Auguste Bravais found that all regular crystals could be described in terms of only 14 lattice types for the 7 crystal systems. Remember that the environment around any one lattice point is exactly the same as the environment around any other lattice point.

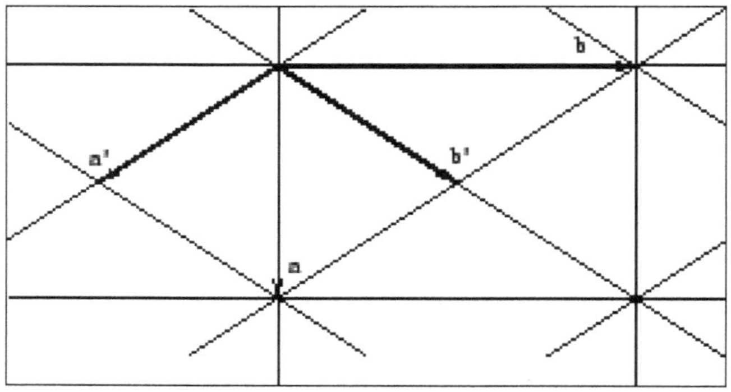

A *c*-centered lattice showing the *a* and *b* lattice vectors and the *a'* and *b'* vectors of the primitive cell.

Non-primitive lattices can have one, two, or three additional lattice points per unit cell. Lattices with one additional lattice point such as an *A*-, *B*-, or *C*-centered cells have an additional lattice point located on the center of some face of the crystal. The additional lattice points in *A*-centered cells appear on the *b-c* faces. Similarly, additional lattice points in *B*-centered cells appear on the *a-c* faces, and lattice points will appear on the *a-b* faces of *C*-centered cells. Lattices with body-centered, *I*-centered, cells have an additional lattice point in the center of the cell. Rhombohedral cells that are based on a hexagonal lattice conventionally have lattice points at (2/3, 1/3, 1/3) and (1/3, 2/3, 2/3). Face-centered, *F*-centered cells have lattice points on all faces.

Generally higher metric symmetry is identified by computer programs. However, these centered lattices can sometimes be identified by a simple examination of the cell parameters. When two cell lengths of a reduced lattice are approximately equal and the two corresponging cell angles are also approximately equal, then the cell is probably centered.

The following table lists the 14 Bravais lattice types. The Bravais symbols are a combination of the crystal system and the lattice designation. Triclinic types begin with the letter a that stands for anorthic from the mineral anorthite a mineral found to have triclinic symmetry. The other lattice types generally begin with the first letter of the crystal system.

Table: Bravais lattice symmetry.

Crystal System	Bravais Lattice	Laue Symmetry
Triclinic	aP	$\overline{1}$ (C_i)
Monoclinic	mP, mS†	2/m(C_{2h})
Orthorhombic	oP, oS*, oF, oI	mmm(D_{2h})
Tetragonal	tP, tI	4/m(C_{4h}), 4/mmm(D_{4h})
Trigonal#		
Hexagonal	hP	$\overline{3}$ (C_{3i}), $\overline{3}$ m(D_{3d})
Rhombohedral	rR	$\overline{3}$ (C_{3i}), $\overline{3}$ m(D_{3d})

Hexagonal	hP	6/m(C_{6h}), 6/mmm(D_{6h})
Cubic	cP, cF, cI	m$\bar{3}$ (T_h), m$\bar{3}$m(O_h)

† The *S* symbol for monoclinic lattices represents a lattice with *A*, *C*, or *I* centering (*b*-unique) or *A*, *B*, or *I* centering (*c*-unique).

* The *S* symbol for orthorhombic lattices stands for any of the three side-centered lattice types, *A*, *B*, or *C*.

Since *P* trigonal lattice and a *P* hexagonal lattice are identical in appearance, these two systems are considered to make up only one Bravais lattice type.

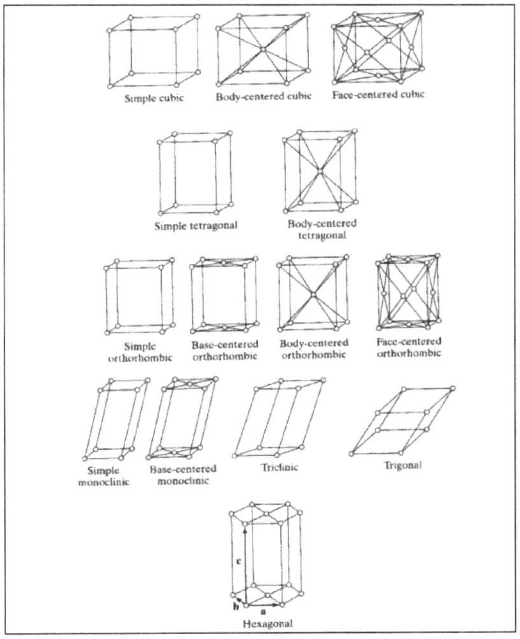

Bravais Lattices.

Rotational Symmetry

There are two basic types of rotational symmetry operations. *Proper* rotations move an object, but do not change the handedness of the object. *Improper* rotations include a proper rotation as well as a component that inverts the handedness of the object.

Proper Rotations

An *n*-fold (C_n) proper rotation operation represents a counter-clockwise movement of (360/n)° around an axis through the object. If an *n*-fold rotation operation is repeated *n* times, then the object returns to its original position. Crystals with a periodic lattice can only have axes with 1-, 2-, 3-, 4-, and 6-fold symmetry axes. The following description of rotational symmetry operations is similar to that given by Prof. Stephen Nelson.

- 1-fold rotation: A 1(*E*)-fold rotation operation implies either a 0° rotation or a 360° rotation, and is referred to as the *identity* operation.

- 2-fold rotation: A 2-fold (*C2*) rotation operation moves the object by $(360/2)° = 180°$. The symbol used to designate a 2-fold axis is a solid oval.

- 3-fold rotation: A 3-fold (*C3*) rotation operation moves the object by $(360/3)° = 120°$. The symbol used to designate a 3-fold axis is a solid equilateral triangle.

- 4-fold rotation: A 4-fold (*C4*) rotation operation moves the object by $(360/4)° = 90°$. The symbol used to designate a 4-fold axis is a solid square.

- 6-fold rotation: A 6-fold (*C6*) rotation operation moves the object by $(360/6)° = 60°$. The symbol used to designate a 6-fold axis is a solid hexagon.

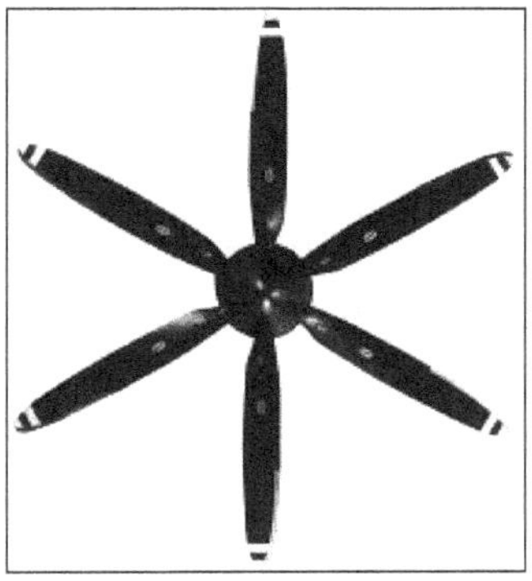

Improper Rotations

An improper rotation may be thought of as occurring in two parts, first a proper rotation is performed, followed by an inversion through a particular point on the rotation axis. In the H-M nomenclature, improper rotations are sometimes called roto-inversions. In the Schönflies scheme, improper rotations are roto-reflection axes because they are a rotation followed by a reflection perpendicular to the rotation axis. Improper rotations are designated by the symbol $\bar{n}$, where n represents the type of proper rotation component of the operation. As in the proper rotation operations, only $\bar{1}$ ($i = S_2$), $m = \bar{2}$ ($\sigma = S_1$), $\bar{3}$ (S_6), $\bar{4}$ (S_4), and $\bar{6}$ (S_3) improper rotations are commonly observed in crystals. These axes are pronounced as 3 bar in the United States and bar 3 in many European countries. Thus 3 in H-M is equivalent to S_6 in Schönflies.

Note that it is not necessary for either the rotation operation or the inversion center to exist as an operation of the group for the improper rotation axis to exist, e.g. the $\bar{4}$ (S_4) operation contains neither a 4-fold rotation axis (C_4) nor an inversion center.

- $\bar{3}$ Roto-inversion: This operation involves a rotation by (360/3) ° followed by an inversion through the center of the object. The symbol is a filled triangle with an open circle in the middle. This is the only improper rotation that also includes the proper rotation axis and an inversion center.
- $\bar{4}$ Roto-inversion: This operation involves a rotation by (360/4) ° followed by an inversion through the center of the object. The symbol is an open 4-sided diamond with an filled oval in the middle.
- $\bar{6}$ Roto-inversion: This operation involves a rotation by (360/6) ° followed by an inversion through the center of the object. The symbol is an open hexagon with an filled triangle inside.

Stereographic Projections

Drawing the three-dimensional symmetry operations on a two-dimensional surface such as this page has been a difficult problem. One way to overcome this problem is through the use of a *stereographic projection.* Such figures are also effective in describing the angular relations among the faces of a crystal.

To construct a stereographic projection, imagine that the object with a given symmetry or surface is at the center of a sphere. Consider the sphere to have a polar axis that is bisected by an equatorial plane. Project features of interest on the object from the center, out to the surface of the sphere. Then project the points on the surface of the sphere through the equatorial plane to the point where the polar axis intersects the sphere in the opposite hemisphere. The stereographic projection is then given by the equatorial plane and all intersections of the plane by the projected points. If the projection point started in the northern hemisphere then its projection onto the equatorial plane is represented as a plus. Points originating in the southern hemisphere are denoted with a circle. Sometimes points generated by improper symmetry operations are also denoted with a comma to indicate opposite handedness.

The unit cell axis with highest symmetry is usually selected as the polar axis. Rotation axes not in the equatorial plane are drawn with the symbol representing the type of axis at the projection point on the equatorial plane. Rotation axes in the equatorial plane are drawn outside of the projection terminating in arrows. Mirror planes are drawn as thickened lines. Inversion centers are drawn as open circles in the center of the polar axis.

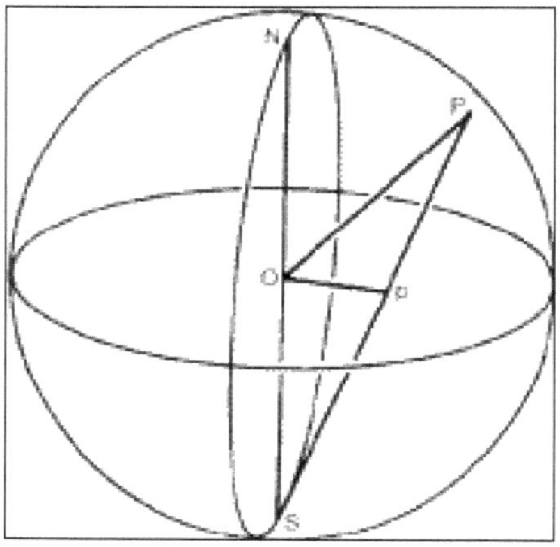

Stereographic projection.

Crystallographic Point Groups

Symmetry operations can be combined to generate other symmetry operations. When these operations are written mathematically, the operations are applied in a right-to-left order. Thus in the expression below, the object x is operated on first by C and then by B. The operation symbol * only indicates that some transformation is being applied to the object.

$$A * x = B * C * x$$

These operations can be thus combined to form a *group* of symmetry operations. These groups of operations are called *point groups* because the symmetry elements of these operations all pass through a single point of the object. In the study of groups, mathematicians have found that such groups always have the following properties.

1. Identity: One operation of the group must exist that when operated either before or after any other group operation produces the same transformation as that of the other group operation. Sometimes the identity operation is called the "do nothing" operation. ($A * E = E * A = A$, E is the identity operation).

2. Inverse: For each operation of the group there must exist a second operation that when combined with the first operation produces the identity operation. The inverse operation cancels the effect of any operation. The identity operation is also its own inverse. ($A * B = B * A = E$, E is the identity, B is the inverse of A).

3. Associativity: The order of combining the operations does not matter. [$A * (B * C) = (A * B) * C$].

In addition to these properties, all crystallographic symmetry groups possess the next property.

4. Closure: When any two operations of the group are combined, then the resultant operation must be a member of the group. ($A = B * C$, A, B, C are operations of the group).

Finally, some groups also exhibit the property of *commutativity*. The order of operations in commutative groups does not matter. $(A * B = B * A)$.

When the proper and improper rotation operations described above are combined following the rules of groups, they yield a total of 32 unique crystallographic point groups. These groups are listed in the following table. The centrosymmetric point groups are shown in bold.

Table: Crystallographic point groups.

System	Essential	Point
	Symmetry	Groups
Triclinic	none	1, $\bar{1}$
Monoclinic	2 or m	2, m, 2/m
Orthorhombic	222 or mm2 *	222, mm2, mmm
Tetragonal	4 or $\bar{4}$	4, 422, $\bar{4}$, 4/m, 4mm, $\bar{4}$2m, 4/mmm
Trigonal	3 or 3	3, $\bar{3}$, 32 †, 3m †, $\bar{3}$m²
Hexagonal	6 or $\bar{6}$	6, 622, $\bar{6}$, 6/m, 6mm, $\bar{6}$2m, 6/mmm
Cubic	23	23, 2/m$\bar{3}$, 432, $\bar{4}$3m, 4/m$\bar{3}$2/m = m$\bar{3}$m

* The symbol *mm*2 also represents the point groups 2*mm* and *m*2*m*.

† These point groups represent sets of groups, e.g., 32 represents 321 and 312

By convention the following rules have been adopted to describe point groups. When a rotation axis is followed by a slash and an m, then this mirror is perpendicular to the rotation axis. For orthorhombic systems the three characters describe the symmetry along the three axes, a, b, and c, respectively. For tetragonal, trigonal, and hexagonal type cells, the c axis is unique, and the first symbol in the point group shows the symmetry along the unique axis. In tetragonal systems, the second symbol shows the symmetry along the [100] and [010] directions and the third symbol shows the symmetry along the [110] and [1$\bar{1}$0] directions. In trigonal and hexagonal cells, the second symbol shows the symmetry along [100], [010] and [$\bar{1}\bar{1}$0], and the third symbol shows symmetry along [210], [120], and [$\bar{1}\bar{2}$0]. In rhombohedral systems on rhombohedral axes, the first symbol shows symmetry along [111], and the second symbol shows symmetry along [1$\bar{1}$0], [01$\bar{1}$], and [$\bar{1}$01]. Cubic symbols show [100], [010], [001] in the first symbol, [111], [1$\bar{1}\bar{1}$], [$\bar{1}$1$\bar{1}$], [$\bar{1}\bar{1}$1] in the second symbol and [1$\bar{1}$0], [110], [01$\bar{1}$], [011], [$\bar{1}$01], and [101] in the third symbol.

The following examples demonstrate how stereographic projections can help understand point groups.

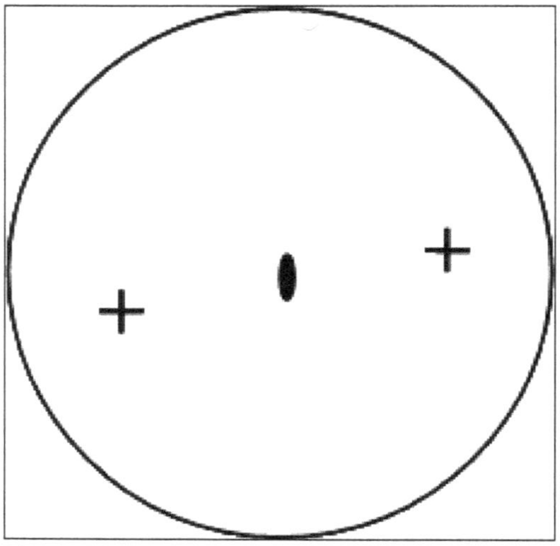

Stereographic projection of 2-fold rotation axis. The cross in the upper, right side of the drawing is assigned the coodinates (x, y, z), the identity operation. If the 2-fold symmetry element is parallel with the b axis then the cross in the lower, left side of the drawing is assigned the coordinates (-x, y, -z).

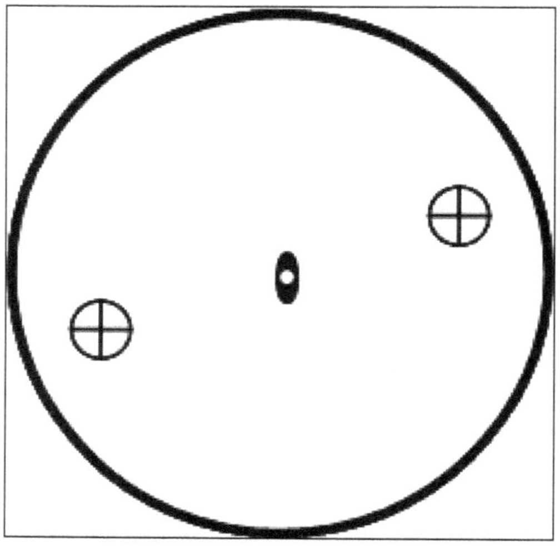

Stereographic projection of 2/m rotation.

In addition to the points in the previous figure, this projection shows a circle in the lower left region with the coordinates $(-x,-y,-z)$ that is a spot generated by the inversion center. Note that the combination of the 2-fold rotation axis and the inversion center leads to another operation--a mirror plane normal to the 2-fold. The relative coordinates of a point related by the mirror in *b* are $(x,-y,z)$. The generation of a mirror by adding an inversion center to a 2-fold axis is an example of the closure property of the group. The mirror is shown by the strong outer circle of the projection.

The other point groups and their symmetry-related coordinates can be derived in a similar manner to that shown above. All 32 crystallographic point groups are shown in the stereographic projections below.

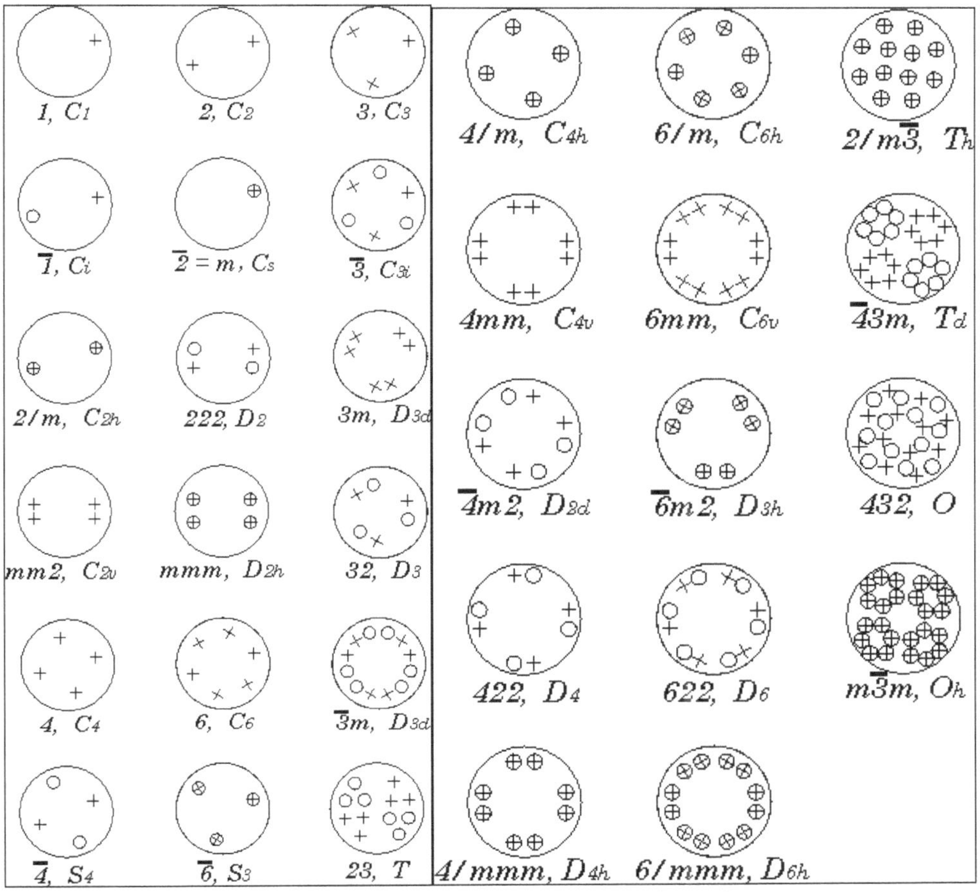

Stereographic projections of the 32 crystallographic point groups.

Micro-translations

Rotational symmetry operations can be combined with translations of part of the unit cell creating entirely new symmetry operations. Proper rotations combined with translations give rise to operations described as *screw* axes. Mirror planes that are combined with translations give rise to *glide* plane operations.

The symbol for a screw axis is n_m where n indicates the type of rotation and the translation is (m/n) of the unit cell. A 2_1 operation is a rotation followed by a translation of 1/2 of the cell parallel with the rotation axis. There are a total of 11 types of screw axes, $2_1, 2_1, 3_1, 3_2, 4_1, 4_2, 4_3, 6_1, 6_2, 6_3, 6_4$, and 6_5. Some pairs of screw axes rotate the objects in opposite directions. The opposite pairs are the 3_1 and 3_2, the 4_1 and 4_3, the 6_1 and 6_5, and the 6_2 and 6_4.

Glide operations occur when a mirror operation is followed by a translation. The translation directions either are parallel with a unit cell direction or are parallel with a combination of cell directions. Glide planes are described by the direction of the translation. The direction of the mirror plane is either explicitly given or is implied for example in the space group symbol. Glide plane operations exist in all three directions and in pairs of directions. Glides that translate by half of the cell in two different directions are called *n* glides. An object undergoes an *n* glide operation reflecting in the (001) plane would be translated by $(a+b)/2$.

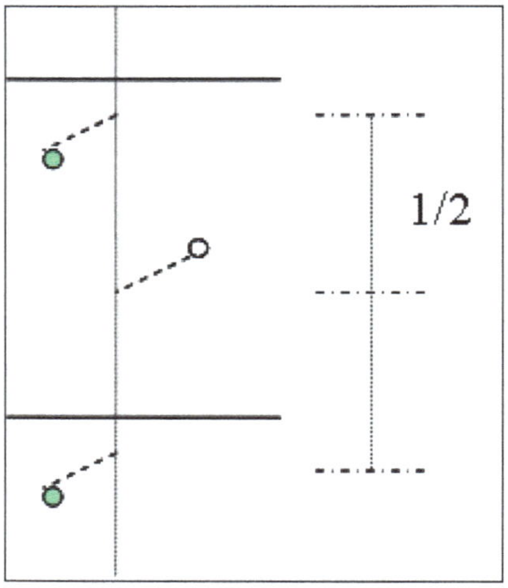

An example of an *a*, *b*, or *c* glide plane.

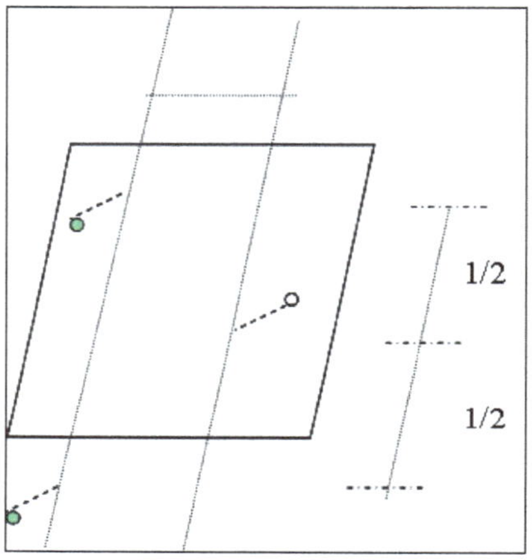

An example of an *n* glide plane.

There is one additional type of glide plane, the diamond glide, *d*. It occurs only in space groups with face- or body-centered cells, and is characterized by a translation of $(\pm a \pm b)/4$, $(\pm b \pm c)/4$, $(\pm c \pm a)/4$, or similar translations. As the denominator implies, 4 consecutive *d* glides are required to return an object to a lattice-translated version of itself.

Symmetry of the Reciprocal Lattice

The real cell parameters are determined by the relative positions of the reciprocal lattice points. Actually the reciprocal cell parameters are determined during a process known as *indexing* the diffraction pattern. From the reciprocal cell parameters the real cell parameters are then calculated according to the relations below.

$$a = b^* \times c^* = (b^* c^* \sin \alpha^*) / V^*$$

$$b = c^* \times a^* = (c^* a^* \sin \beta^*) / V^*$$

$$c = a^* \times b^* = (a^* b^* \sin \gamma^*) / V^*$$

$$V^* = 1/V = a^*b^*c^*\left(1 - \cos^2\alpha^* - \cos^2\beta^* - \cos^2\gamma^* \;+\; 2\ \cos\alpha^*\ \cos\beta^*\cos\ \gamma^*\right)$$

$$\cos\alpha = \left(\cos\ \beta^*\cos\gamma^* - \cos\alpha\right) / \left(\sin\beta^*\sin\ \gamma^*\right)$$

$$\cos\beta = \left(\cos\gamma^*\cos\alpha^* - \cos\beta\right) / \left(\sin\gamma^*\sin\alpha^*\right)$$

$$\cos\gamma = \left(\cos\alpha^*\cos\beta^* - \cos\gamma\right) / \left(\sin\alpha^*\sin\beta^*\right)$$

Laue Class

The relative intensities in a diffraction pattern are dependent on the electron density distribution of the sample. This electron density distribution must follow the symmetry of the crystal itself. This symmetry is called the *Laue class*. The Laue class for a sample is described as one of the 11 centrosymmetric point groups. Note that the appropriate centrosymmetric point group or Laue class for a sample can be identified by adding a center of symmetry to the point group operations of the particular crystallographic point group of the sample.

Friedel's Law

The additional center of symmetry is due, at least approximately, to the fact that diffraction or interference effects are inherently centrosymmetric. The intensity of the $(h\ k\ l)$ point in the reciprocal lattice comes from scattering of the electron density that is parallel to the $(h\ k\ l)$ planes in the crystal. Similarly, the intensity of the $(\bar{h}\ \bar{k}\ \bar{l})$ point comes from the electron density in planes parallel with the $(\bar{h}\ \bar{k}\ \bar{l})$ planes in the crystal. But since the $(h\ k\ l)$ planes and the $(\bar{h}\ \bar{k}\ \bar{l})$ planes are simply opposite in direction, then the intensities of the $(h\ k\ l)$ and $(\bar{h}\ \bar{k}\ \bar{l})$ points in the reciprocal lattice should be, at least approximately the same. This equality relationship between the intensites of $(h\ k\ l)$ and $(\bar{h}\ \bar{k}\ \bar{l})$ is called *Friedel's law*.

For the intensity data from a chiral compound, Friedel's law can be broken by the anomalous scattering of heavy atoms. In these data sets the reciprocal lattice has the same symmetry as the symmetry of the point group of the crystal. Thus if the point group symmetry of the crystal is shown to be 222 then the intensities would exhibit 222 symmetry. The anomalous scattering of heavy atoms is not a strong effect, so the intensities will still approximately exhibit Friedel's law.

Symmetry-equivalent Intensities

One simple way to determine which data should have equivalent intensities based on symmetry is to consider the stereographic projection of the point group of the sample. From a simple examination of the stereographic projection, you can determine the symmetry-related (x, y, z) coordinates. Simply convert the x coordinates to *h*, the y coordinates to *k*, and the z coordinates to *l* values. The symmetry-equivalent intensities are then determined.

A monoclinic crystal, has the Laue symmetry of $2/m$. The equivalent coordinates, assuming a *b*-unique axis, are given as (x, y, z), (-x, y, -z), (-x, -y, -z), and (x, -y, z). Thus the intensities of the $(h\ k\ l)$, $(\bar{h}\ k\ \bar{l})$, $(\bar{h}\ \bar{k}\ \bar{l}$,and $(h\ \bar{k}\ l)$ lattice points should have equivalent values. Note that this also means that the intensities of the $(h\,k\,\bar{l})$, $(\bar{h}\,k\,l)$, $(\bar{h}\,\bar{k}\,l)$, and $(h\,\bar{k}\,\bar{l})$ should also be equivalent to each other but are not necessarily equivalent to $(h\,k\,l)$, etc.

If a monoclinic compound is chiral then the intensities would be expected to have only 2 point group symmetry. Thus the intensities of the $(h\,k\,l)$ data are equivalent to the intensities of the $(\bar{h}\,k\,\bar{l})$ data. Similarly, $I(h\,k\,\bar{l}) = I(\bar{h}\,k\,l); I(h\,\bar{k}\,l) = I(\bar{h}\,\bar{k}\,\bar{l})$; and $I(h\,\bar{k}\,\bar{l}) = I(\bar{h}\,\bar{k}\,l)$.

If a crystal happens to have all three cell angles = 90.0° within experimental error then most workers would guess that the sample had orthorhombic symmetry. In most cases this guess would be correct, but not in all cases. If $I(h\,k\,l) = I(\bar{h}\,k\,\bar{l})$ and $I(h\,\bar{k}\,\bar{l}) = I(\bar{h}\,\bar{k}\,l)$, but $I(h\,k\,l) \neq I(h\,\bar{k}\,\bar{l})$, then the sample has monoclinic not orthorhombic Laue symmetry. The symmetry of the Laue class is dictated by the symmetry of the *reciprocal lattice intensities* not the apparent symmetry of the cell parameters. The cell parameters only dictate the *possible* highest symmetry of the sample.

Systematic Absences

Some symmetry operations can be readily identified by specific information in the intensities of the diffraction pattern. In particular, cell centering, screw axes, and glide plane operations can be identified by the fact that they cause certain groups of diffraction points to be systematically absent. These symmetry operations all include a micro-translation.

Consider a data set that has a *c* glide operation reflecting in the plane normal to the *b* axis. The symmetry operations could be (x, y, z) and (x, -y, z+1/2). If there are N atoms in the unit cell, then there are N/2 unique atoms. The summations below are over the *j* atoms and run from 1 to N/2.

$$\boldsymbol{F}(hkl) = \Sigma f_j \exp\, 2\pi i\left(hx_j + ky_j + lz_j\right) +$$
$$\Sigma f_j \exp\, 2\pi i\left[hx_j - ky_j + l\left(1/2\ + z_j\right)\right]$$

Consider the data with k = 0. For these data the structure factors become:

$$\boldsymbol{F}(h0l) = \Sigma f_j \exp\, 2\pi i\left(hx_j + lz_j\right) +$$
$$\Sigma f_j \exp\, 2\pi i\left(hx_j + lz_j\right)\ \exp 2\pi i\left(l/2\right)$$

$$\boldsymbol{F}(h0l)\ =\ \Sigma f_j \exp\, 2\pi i\left(hx_j + lz_j\right)\ \left[1 + \exp \pi i l\right]$$

If l is an odd integer, then $\exp \pi i l = -1$ and $\boldsymbol{F}(h0l) = 0$. If l is an even integer, then $\boldsymbol{F}(h0l)$ is probably not (but could accidentally be) 0. The systematic absence conditions for other symmetry operations can be derived in a similar manner as was done above.

A table of these absence conditions is shown below. The reflection condition indicates the data that may be present, other data with the same conditions would be systematically absent. The character n can be any integer. Thus the condition that for $hkl, k + l = 2n + 1$, indicates that for the general class of peaks hkl that the sum of $k + l$ should be positive for these peaks to have a measurable intensity, and that those peaks with $k + l$ negative must not have a measurable intensity if the *A* centering symmetry is present in the symmetry of the data set.

It is always best to check for systematic absences, and hence for translation-containing symmetry operations, in the following order cell centering operations, then glide planes, and finally screw axes. This order is important because higher symmetry operations such as cell centering can mask lower symmetry operations such as glide planes and screw axes. Also, this order works from conditions that have the largest number of possible absences to the conditions with the least possible absences for a given sample.

Note that if a crystal exhibits the systematic absence condition for *C* cell centering then the fact that for $0kl, k = 2n$ does not necessarily indicate a *b* glide reflecting in *a*. This condition gives no new information, because it is a part of the cell centering condition.

Table: Systematically Absent Reflection Conditions.

Symmetry Element	Types	Reflection Condition
A centered	hkl	k + l = 2n
B centered		h + l = 2n
C centered		h + k = 2n
F centered		k + l = 2n, h + l = 2n, h + k = 2n
I centered		h + k + l = 2n
R (obverse)		-h + k + l = 3n
R (reverse)		h - k + l = 3n
Glide reflecting in a	0kl	
b glide		k = 2n
c glide		l = 2n
n glide		k + l = 2n
d glide		k + l = 4n
Glide reflecting in b	h0l	
a glide		h = 2n
c glide		l = 2n
n glide		h + l = 2n
d glide		h + l = 4n
Glide reflecting in c	hk0	
b glide		k = 2n
a glide		h = 2n
n glide		k + h = 2n
d glide		k + h = 4n
Glide reflecting in (110)	hhl	
b glide		h = 2n
n glide		h + l = 2n
d glide		h + k + l = 4n
Screw $\parallel$	h00	
2_1, 4_2		h = 2n
4_1, 4_3		h = 4n
Screw $\parallel$	0k0	

2_1, 4_2		k = 2n
4_1, 4_3		k = 4n
Screw ||	ool	
2_1, 4_2, 6_3		l = 2n
3_1, 3_2, 6_2, 6_4		l = 3n
4_1, 4_3		l = 4n
6_1, 6_5		l = 6n
Screw ||	hho	
2_1		h = 2n

Space Groups

When the 7 crystal systems are combined with the 14 Bravais lattices, the 32 point groups, screw axes, and glide planes, Arthur Schönflies, Evgraph S. Federov, and H. Hilton were able to describe the 230 unique space groups. A space group is a group of symmetry operations that are combined to describe the symmetry of a region of 3-dimensional space, the unit cell. In point groups, the symmetry elements all pass through one point in the object. In space groups, symmetry elements need not intersect at a single point, although some operations may coincidentally intersect at various points in the cell.

Chiral compounds that are prepared as a single enantiomer can crystallize in only a subset of 65 space groups that do *not* have improper rotations such as centers of symmetry, mirror planes, glides, or $\bar{3}$, $\bar{4}$, or $\bar{6}$ axes.

A space group is designated by a capital letter identifying the lattice type (*P*, *A*, *F*, etc.) followed by the point group symbol in which the rotation and reflection elements are extended to include screw axes and glide planes. Note that the point group symmetry for a given space group can be determined by removing the cell centering symbol of the space group and replacing all screw axes by similar rotation axes and replacing all glide planes with mirror planes. The point group symmetry for a space group describes the true symmetry of its reciprocal lattice.

The Bilbao Crystallographic Server gives a description of all space groups and their symmetry generators, general positions, Wyckoff positions, related subgroups and supergroups, and normalizers. M. W. Meier has prepared an interesting set of notes on space group patterns that illustrates the various combinations of point group symmetry elements into space groups. The following table lists the unique space groups separated by crystal system and Laue class.

Table: Space groups based on crystal system and Laue class.

Crystal System	Laue Class	Space Group
Triclinic	$\bar{1}$	P1, P$\bar{1}$
Monoclinic	2/m	P2, P2_1, C2, Pm, Pc, Cm, Cc, P2/m, P2_1/m, C2/m, P2/c, P2_1/c, C2/c
Orthorhombic	mmm	P222, P22$2_1$, P$2_1$$2_1$2, P$2_1$$2_1$$2_1$, C22$2_1$, C222, F222, I222, I$2_1$$2_1$$2_1$, Pmm2, Pmc$2_1$, Pcc2, Pma2, Pca$2_1$, Pnc2, Pmn$2_1$, Pba2, Pna$2_1$, Pnn2, Cmm2, Cmc$2_1$, Ccc2, Amm2, Aem2, Ama2, Aea2, Fmm2, Fdd2, Imm2, Iba2, Ima2, Pmmm, Pnnn, Pccm, Pban, Pmma, Pnna, Pmna, Pcca, Pbam, Pccn, Pbcm, Pnnm, Pmmn, Pbcn, Pbca, Pnma, Cmcm, Cmce, Cmmm, Cccm, Cmme, Ccce, Fmmm, Fddd, Immm, Ibam, Ibca, Imma

Tetragonal	4/m	$P4, P4_1, P4_2, P4_3, I4, I4_1, P\bar{4}, I\bar{4}, P4/m, P4_2/m, P4/n, P4_2/n, I4/m, I4_1/a$
Tetragonal	4/mmm	$P422, P42_12, P4_122, P4_12_12, P4_222, P4_22_12, P4_322, P4_32_12, I422, I4_122, P4mm, P4bm, P4_2cm, P4_2nm, P4cc, P4nc, P4_2mc, P4_2bc, I4mm, I4cm, I4_1md, I4_1cd, P\bar{4}2m, P\bar{4}2c, P\bar{4}2_1m, P\bar{4}2_1c, P\bar{4}m2, P\bar{4}c2, P\bar{4}b2, P\bar{4}n2, I\bar{4}m2, I\bar{4}c2, I\bar{4}2m, I\bar{4}2d, P4/mmm, P4/mcc, P4/nbm, P4/nnc, P4/mbm, P4/mnc, P4/nmm, P4/mcc, P4_2/mmc, P4_2/mcm, P4_2/nbc, P4_2/nnm, P4_2/mbc, P4_2/mnm, P4_2/nmc, P4_2/ncm, I4/mmm, I4/mcm, I4/amd, I4/_1acd$
Trigonal	$\bar{3}$	$P3, P3_1, P3_2, R3, P\bar{3}, R\bar{3}$,
Trigonal	$\bar{3}$m	$P312, P321, P3_112, P3_121, P3_212, P3_221, R32, P3m1, P31m, P3c1, P31c, R3m, R3c, P\bar{3}1m, P\bar{3}1c, P\bar{3}m1, P\bar{3}c1, R\bar{3}m, R\bar{3}c$
Hexgonal	6/m	$P6, P6_1, P6_5, P6_2, P6_4, P6_3, P\bar{6}, P6/m, P6_3/m$
Hexagonal	6/mmm	$P622, P6_122, P6_522, P6_222, P6_422, P6_322, P6mm, P6cc, P6_3cm, P6_3mc, P\bar{6}m2, P\bar{6}c2, P\bar{6}2m, P\bar{6}2c, P6/mmm, P6/mcc, P6_3/mcm, P6_3/mmc$
Cubic	m$\bar{3}$	$P23, F23, I23, P2_13, I2_13, Pm\bar{3}, Pn\bar{3}, Fm\bar{3}, Fd\bar{3}, Im\bar{3}, Pa\bar{3}, Ia\bar{3}$
Cubic	m$\bar{3}$m	$P432, P4_232, F432, F4_132, I432, P4_332, P4_132, I4_132, P\bar{4}3m, F\bar{4}3m, I\bar{4}3m, P\bar{4}3n, F\bar{4}3c, I\bar{4}3d, Pm\bar{3}m, Pn\bar{3}n, Pm\bar{3}n, Pn\bar{3}m, Fm\bar{3}m, Fm\bar{3}c, Fd\bar{3}m, Fd\bar{3}c, Im\bar{3}m, Ia\bar{3}d$

The space groups in bold are centrosymmetric.

The previous table lists the mathematically-unique space groups. In addition to these there are many non-standard space groups. For example, the space groups $P2_1/a$ and $P2_1/n$ are variants of the space group $P2_1/c$. Some non-standard space groups are reported in a publication so that the cell parameters of several related compounds can be more easily compared. For example, if the main compound in a paper is found to be in the monoclinic space group *C2/c*, but a related compound is found to be triclinic, some authors would determine the structure of this second compound in the non-standard space group $C\bar{1}$ if the cell parameters in $C\bar{1}$ were similar to the cell parameters of the main compound. Finally some authors would publish structures in non-standard space groups because the basis vectors of the related cells would produce cell angles that were closer to 90 °. For example, structures are often published in *Pn* rather than *Pc* or in *I2/a* rather than *C2/c* because the β angle would be much closer to 90°. When β angles are closer to 90°, the refinement of the x and z coordinates of the atoms show much less correlation, and converge to form chemically-reasonable structures more quickly.

The following tables list the graphical and typed symbols used to describe symmetry operations in the International Tables for Crystallography, Vol A. Please note that the symbol ⊥ is used to indicate that the operation is perpendicular (normal) to the page.

Table: Symmetry planes normal to the plane of projection.

Symmetry plane	Graphical symbol	Translation	Symbol
Reflection plane	————	None	m
Glide plane	- - - - - - - - - -	1/2 along line	a, b, or c
Glide plane	....	1/2 normal to plane	a, b, or c

Double glide plane	· — ·· — ··	1/2 along line & 1/2 normal to plane	e
Diagonal glide plane	— · — · —	1/2 along line & 1/2 normal to plane	n
Diamond glide plane	— · — · ←	1/4 along line & 1/4 normal to plane	d

Table: Symmetry planes parallel to the plane of projection.

Symmetry plane	Graphical symbol	Translation	Symbol
Reflection plane		None	m
Glide plane		1/2 along arrow	a, b, or c
Double glide plane		1/2 along either arrow	e
Diagonal glide plane		1/2 along the arrow	n
Diamond glide plane	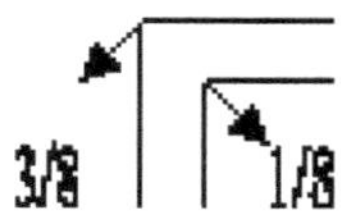	1/8 or 3/8 along the arrows	d

Table: Symmetry Axes.

Symmetry Element	Graphical Symbol	Translation	Symbol
Identity	None	None	1
2-fold ⊥ page	●	None	2
2-fold in page		None	2
2 sub 1 ⊥ page		1/2	2_1
2 sub 1 in page		1/2	2_1
3-fold		None	3
3 sub 1		1/3	3_1

3 sub 2		2/3	3_2
4-fold		None	4
4 sub 1		1/4	4_1
4 sub 2		1/2	4_2
4 sub 3		3/4	4_3
6-fold		None	6
6 sub 1		1/6	6_1
6 sub 2		1/3	6_2
6 sub 3		1/2	6_3
6 sub 4		2/3	6_4
6 sub 5		5/6	6_5
Inversion		None	1
3 bar		None	3
4 bar		None	4
6 bar		None	6 = 3/m
2-fold and inversion		None	2/m

2 sub 1 and inversion		None	$2_1/m$
4-fold and inversion		None	4/m
4 sub 2 and inversion		None	$4_2/m$
6-fold and inversion		None	6/m
6 sub 3 and inversion		None	$6_3/m$

Determining the Space Group of a Material

Identifying the proper space group for a given sample and its diffraction pattern is often straight forward. The Laue class is determined first. Then the possible systematic absences are determined to identify the appropriate cell centering conditions, glide planes, and screw axes, if any. From the Laue class and the symmetry operations identified by the systematic absences, the choice of space groups is usually limited one unique or to a very few.

If the space group is not uniquely determined then the structure solution and refinement steps are tried with the different possible space groups until the structure determination is completed. Try the space groups beginning with the highest symmetry first. If the chosen space group is not the highest symmetry space group, then be sure to check for additional symmetry elements in the structure using either the PLATON program or the check cif program.

Additional information about the compound is often used to reduce the choice of space group to a single one. If the compound is chiral and is expressed as a single enantiomorph, for example, then the compound can only crystallize in one of the 65 chiral space groups.

Merging Data to Determine Laue Class

The Laue class is usually found by calculating the merging *Rint* of the symmetry-equivalent data for the given Laue class. An appropriate *Rint* for identifying the Laue class would be a value & 0.06. Values larger than this usually suggests that the sample crystallized in a lower symmetry Laue group.

Symmetry-equivalent intensity data are merged using the following relationship:

$$F^2 = \sum \omega_j F_j^2 / \sum \omega_j$$

Where the summations are over the set of symmetry-equivalent data. Sometimes *outliers* are either removed from the merged set or are given lower weights to reduce their effects on the subsequent refinement.

$$Rint = \Sigma\left[\Sigma\left|F_j^2 - \langle F^2 \rangle\right|\right] / \Sigma\left[\left(\Sigma F_j^2\right)/n\right]$$

Where the inner sums are over the symmetry-equivalent reflections and the outer sums are over the unique *hkl* data. The term n is the number of equivalent data for a given *hkl* being merged.

Tests for a Center of Symmetry

If the compound is not a single enantiomorph, then statistical tests for a center of symmetry can be performed.

Probability distributions for centric and acentric unit cells have been derived and are given below. These derivations assume that the electron density is randomly and uniformly distributed throughout the unit cell.

$$P_{-1}(|F|) = \left[(2)^{1/2} / (\pi S)^{1/2}\right] \exp\left(-|F|^2 / 2S\right) (\text{centric})$$

$$P_1(|F|) = (2|F| / S) \exp\left(-|F|^2 / S\right) \quad (\text{acentric})$$

Where,

$$S = \Sigma f_j$$

S is a function of the scattering factors for the atoms and thus will decrease for increasing sin (θ / λ). To simplify this functional dependence of the distributions on θ, most statistical tests utilize normalized structure factors, E_{hkl}.

$$E_{hkl}^{\ 2} = F_{hkl}^{\ 2} / \left(\varepsilon \Sigma f_j^{\ 2}\right)$$

where $f_j = f_j^{\,o} \exp\left(-B \sin^2 \theta / \lambda_2\right)$ is the scattering factor for the jth atom and ε is an integer, 1 or greater, that corrects for the fact that some classes of reflections have expectation values that are less than $\Sigma f_j^{\ 2}$ by an integer amount.

Since $\langle |E_{\mathbf{h}}|^2 \rangle = 1 = S$, the distributions become:

$$P_{-1}(|E|) = \left[(2 / \pi)^{1/2}\right] exp\left(-|E|^2 / 2\right) \quad (\text{centric})$$

$$P_1(|E|) = 2\,|E|\, exp\left(-|E|^2\right) \quad (\text{acentric})$$

These two distributions are completely independent of the variation of θ and are significantly different from each other. The differences in the two distributions indicate that centric unit cells should have more reflections with very strong and very weak values compared to acentric unit cells. Acentric unit cells typically tend to have more even distributions of data with significantly fewer weak data than centric cells. Comparing the distribution of a measured data set with the two theoretical distributions could establish the presence or absence of a center of symmetry in the crystal under consideration.

Mean values for a variety of functions of $|E|$ may be estimated from the two theoretical distributions. These values are presented in the following Table. Most computer programs that test for centric unit cells calculate the mean values for these functions of E and compare these experimental values with the theoretical values.

Table: Theoretical values for selected functions of |E|.

	Centric	Acentric
$<\|E\|^2>$	1.000	1.000
$<\|E^2-1\|>$	0.968	0.736
$<\|E\|>$	0.798	0.886

Three caveats must be remembered when using these tests. First, the intensity data must be relatively strong. Weak data sets, having $F^2/\sigma < 6.0$, have too many weak data to give reliable results for this test. Second, these theoretical values were derived assuming a uniform distribution of electron density in the unit cell. The presence of heavy atoms or of an unusual distribution of atoms tends to make the experimental values assume a centric distribution. When the tests indicate an acentric unit cell, the tests are likely to be correct. When the tests suggest a centric unit cell, the tests may be correct or may simply indicate a skewed distribution of electron density in the unit cell. Finally, if the tests indicate a hyper acentric distribution, e.g. $<|E^2-1|>< 0.6$, then it is likely that the sample is twinned.

Polar Space Groups

Some space groups do not contain symmetry operators that fix the origin of the unit cell in one or more directions. These space groups can be identified by the examining the symmetry-equivalent positions for the space group. For any of the three coordinate directions there are no symmetry operations that invert that coordinate $(x \text{-}\!> \text{-}x)$, $(y \text{-}\!> \text{-}y)$, or $(z \text{-}\!> \text{-}z)$. For these *polar* space groups, the origin must be fixed by the atoms. This may be accomplished either by holding the coordinates in the polar directions fixed for one atom during refinement or by applying a constraint to hold some function, usually the mean value, of all coordinates in the given directions fixed. If the polar axis is established by fixing the coordinate or coordinates of one atom, then selecting the heaviest atom in the structure for this purpose will lead to a more stable refinement.

Inversion of a Crystal Structure

Very few methods of solving a structure are able to properly pick the correct *absolute configuration* of a crystal structure. For these samples, the structure is roughly determined and refined. The configuration is then checked either by means of anomalous scattering or by knowing and checking the configuration of the compound itself. For example, naturally occuring peptides should have most helices with a right-handed twist. There may be a few short helices with a left-handed twist, but the nearly all should have a right-handed twist.

If the structure is determined to have the incorrect handedness, then the structure is simply inverted through some symmetry point such as (0, 0, 0) or (0.5, 0.5, 0.5) and the refinement is completed. The space group must also be changed if the structure crystallizes in one of the 11 enantiomorphous pairs of space groups. If the sample crystallizes in one of 7 special space group, then inversion must be through a special point.

The 230 space groups include 11 space groups that occur as enantiomorphous pairs: $P3_1$, $P3_2$; $P3_112$, $P3_212$; $P3_121$, $P3_221$; $P4_1$, $P4_3$; $P4_122$, $P4_322$; $P4_12_12$, $P4_32_12$; $P6_1$, $P6_5$; $P6_2$, $P6_4$; $P6_122$, $P6_522$; $P6_222$, $P6_422$; and $P4_132$, $P4_332$. When a sample crystallizes in one of these space groups, one of the pair is arbitrarily chosen for the structure solution and at least partial refinement. If it is later determined that the wrong space group of the pair was chosen initially, then the atoms are inverted and the alternate space group of the pair is used in further refinements.

For 7 other space groups inversion must be carried out through some point in the structure other than (0, 0, 0) or (0.5, 0.5, 0.5). This problem was first described in print by Parthe and Gelato and Bernardinelli and Flack. A table listing these seven space groups and appropriate inversion points is listed below.

Table: Space groups with non-origin inversion points.

Space Group	Inversion Point	SHELXTL MOVE*
Fdd2	0.125 0.125 0.500	MOVE 0.25 0.25 1 -1
$I4_1$	0.500 0.250 0.500	MOVE 1 0.5 1 -1
$I4_122$	0.500 0.250 0.125	MOVE 1 0.5 0.25 -1
$I4_1md$	0.500 0.250 0.500	MOVE 1 0.5 1 -1
$I4_1cd$	0.500 0.250 0.500	MOVE 1 0.5 1 -1
$I\bar{4}2d$	0.500 0.250 0.125	MOVE 1 0.5 0.25 -1
$F4_132$	0.125 0.125 0.125	MOVE 0.25 0.25 0.25 -1

* The command in the SHELX refinement program to invert the atoms.

Special Positions

Many space groups contain simple symmetry operations, such as inversion centers, rotation axes, and mirror planes. Usually the locations of these symmetry elements are either fixed by convention, or fixed relative to other symmetry operations in the cell. These special positions are located on points (centers of symmetry), lines (rotations axes) or planes (mirror planes). Also, there are combinations of symmetry operations such as 2/m points or mm lines.

An atom at one of these locations will have fewer symmetry-related positions in the cell than would an atom in a general position. The International Tables, Vol. A lists special positions often called Wyckoff positions as well as the relative number of symmetry-related positions.

Because atoms on some of these special positions are a constant distance from their symmetry-related atoms, these special positions can give rise to systematically absent intensity data. These systematic absences can be derived in a similar manner as the systematic absence conditions for glide planes and screw axes derived above.

Note that these systematic absence conditions apply only to atoms that sit on the respective special positions. This can be a problem when the atoms on a special position are very heavy and the remaining atoms are in general positions are much lighter and are, in general, evenly distributed in the cell. Under these circumstances, the systematic absence due to the atom on the special position could be confused as a general systematic absence, obscuring the space group symmetry. Thus when a structure cannot be solved, and other tricks have been tried, it may be necessary to lower the symmetry to try to solve the structure.

Space Group Names

Although the Hermann-Mauguin names for space groups will always be used in publications and in the cifs accompanying the publications, another set of names for space groups is also included in cif files. This alternate description of space groups was developed by Sid Hall to give a unique name to each space group and space group setting. However, all of the ambiguities in H-M space group names are resolved by looking at either the cell parameters or the symmetry operators.

Crystallography of Functional Materials

Functional materials, which possess specific properties and functions, form the basis for a wide range of technologies, including electronics, information storage and retrieval, communication, biotechnology, aerospace, defense, environmental, energy, medical, and consumer products. A partial list of the special chemical, electrical, optical, and physical properties of these materials includes magnetism, semiconducting behavior, ferroelectricity, piezoelectricity, ferromagnetism, ionic conductivity, superconductivity, photocatalytic behavior, adsorption behavior, and giant magneto-resistance, with resulting specific applications in microelectronic devices and data storage, sensing, energy storage, hydrogen storage, thermoelectric waste heat conversion, solar cells, fuel cells, batteries, and other energy harvesting and conversion applications. Functional materials of interest cover different classes of solid-state materials ranging from semiconductors to polymers, from molecular crystals to nanoparticles. The development and fundamental understanding of the properties of functional materials are central to technological needs, and are at the forefront of materials research.

The field of crystallography provides the foundation of the structure–properties relationship. It bridges the fundamental understanding of materials with their applications. As functional materials play a critical role in our modern society and future economic, physical, and social success, the underlying crystallographic information for these materials is of utmost importance for many research communities.

To date, a number of journals have been devoted to the important subject of functional materials including Journal of Functional Materials, Functional Materials Letters, Advanced Functional

Materials, and Journal of Functional Materials and Devices. However, despite the diverse subjects covered in these journals, there has not been a focus on the importance of the relationships between crystallography and representative functional properties.

Representative Functional Materials

Ge-waveguide Materials

An optical waveguide is a physical structure that guides electromagnetic waves in the optical spectrum. Common types of optical waveguides include optical fiber and rectangular waveguides. Optical waveguides are used as components in integrated optical circuits or as the transmission medium in local and long haul optical communication systems. These waveguides in general are classified according to their geometry (planar, strip, or fiber waveguides), mode structure (single-mode, multi-mode), refractive index distribution (step or gradient index), and material (glass, polymer, semiconductor).

Tensile-strained Ge has been investigated as a possible laser material due to two main factors, (1) nearly-direct band gap and (2) its compatibility with conventional silicon integrated circuit fabrication. However, despite the multitude of benefits, problems remain due to the fact that lasers using tensile-strained Ge show a high threshold current density and short lifetime. To understand the origins of these problems, the authors used the combined techniques of Raman spectroscopy and time-dependent diffraction techniques such as high-resolution transmission electron microscopy (HRTEM). Real-time degradation in tensile-strained Ge was observed, and a network of defects was observed to propagate with a speed and density correlated with the local strain. Therefore, highly tensile-strained Ge lasers (which can lead to the formation of dislocations and stacking faults) indeed have a short lifetime, and they are better suited for low-energy devices than they are for lasers.

Advanced Li-ion Battery Materials

Lithium-ion batteries, invented in the late 1970s and early 1980s and commercialized in the 1990s, generate electrical currents when lithium ions travel between two electrodes. Light but powerful, they have transformed portable electronics. Their use in electric cars, however, is recent. When Tesla Motors introduced one of the first lithium-ion-powered electric cars in 2008, it could go 250 m on a charge, but the vehicle cost over $100,000, in large part because the batteries are expensive . To cut costs, the lithium-ion-powered electric cars made today by companies such as Nissan and GM use small battery packs with a range of less than 100 m. While countless breakthroughs have been announced over the last decade, these advances failed to translate into commercial batteries. One difficulty about developing better batteries is that the Li-ion battery technology is still poorly understood, including the crystallography of the important cathode oxide materials.

Bendersky et al. investigated the crystallography and growth of epitaxial oxide films for an in-depth understanding of the cathode materials used in advanced Li-ion batteries. Details of film growth, including the identity of each of the layer forms, are important to understand the charge and discharge mechanism. High-resolution Transmission Electron Microscopy (HRTEM) is by far the most efficient and sophisticated technique for identifying various layers of film growth and their crystallographic relationships. Bendersky et al. studied a number of Li-containing oxide compositions representing different functions. Results indicated that epitaxial film growth is possible for all films, with the growth following the predicted cube-on-cube orientation relationship between the cubic and pseudo-cubic oxygen sublattices of a film and a substrate. These results provide new insights into electrochemical processes occurring in the bulk and at interfaces.

Luminescence Properties of Borates

Inorganic light emitting diode (LED) materials are one of the major research areas within solid-state luminescence. Of special interest is generating white LED light and one approach to achieve this uses an inorganic phosphor material to convert near-UV radiation, emitted from a LED, to white light. In the paper by Zhao et al., the authors report on the crystal structure of $Na_3Sm(BO_3)_2$ as a potential crystalline host material. Recently, Wang et al. reported and characterized the $Na_3Sm(BO_3)_2$ compound using X-ray powder diffraction, while the work presented in this special volume by Zhao et al. demonstrates the growth of a single crystal of $Na_3Sm(BO_3)_2$, using a high temperature flux method. Zhao et al. studied the atomic structure via single crystal X-ray diffraction and the thermal stability of the compound using TG/DTA. The paper also presents the ultraviolet-visible (UV-vis) absorption spectrum, the excitation and emission spectra, decay time, and location of the $Na_3Sm(BO_3)_2$ phosphor on the Commission International de l'Éclairage (CIE) chromaticity index. The results of their structural and luminescent properties characterization lead the authors to conclude that $Na_3Sm(BO_3)_2$ could potentially be used as an orange–red phosphor under near-UV light excitation.

Luminescence Properties of Borate

Inorganic light emitting diode (LED) materials are one of the major research areas within solid-state luminescence. Of special interest is generating white LED light and one approach to achieve this uses an inorganic phosphor material to convert near-UV radiation, emitted from a LED, to white light. In the paper by Zhao et al., the authors report on the crystal structure of $Na_3Sm(BO_3)_2$ as a potential crystalline host material. Recently, Wang et al. reported and characterized the $Na_3Sm(BO_3)_2$ compound using X-ray powder diffraction, while the work presented in this special volume by Zhao et al. demonstrates the growth of a single crystal of $Na_3Sm(BO_3)_2$, using a high temperature flux method. Zhao et al. studied the atomic structure via single crystal X-ray diffraction and the thermal stability of the compound using TG/DTA. The paper also presents the ultraviolet-visible (UV-vis) absorption spectrum, the excitation and emission spectra, decay time, and location of the $Na_3Sm(BO_3)_2$ phosphor on the Commission International de l'Éclairage (CIE) chromaticity index. The results of their structural and luminescent properties characterization lead the authors to conclude that $Na_3Sm(BO_3)_2$ could potentially be used as an orange–red phosphor under near-UV light excitation.

Layered Magnetic Metal Halides

The crystal structure and magnetic properties of MX_2 and MX_3 compounds where M is a transition metal with partially filled d-shells (M = Ti, V, Cr, Mn, Fe, Co, Ni, Zr, Mo, Tc, Ru, Rh, and Ir) and X is either Cl, Br, or I. When M = Ti, V, Mn, Fe, Co, or Ni the MX_2 compound crystallizes in either the CdI_2 or $CdCl_2$ structure types. However, when M = Zr and X = Cl, the compound crystallizes with the MoS_2 structure type; when X = I, it can crystalize as either the $MoTe_2$ or WTe_2 structure types.

The CdI_2 and $CdCl_2$ structure types have in common edge sharing octahedra, creating triangular nets of cations resulting in MX_2 layers perpendicular to the c-direction, and are distinguished by layering in the c-direction. The MX_3 compounds predominately crystalize in the BiI_3 or $AlCl_3$ structure types and again have edge sharing octahedra, but for these structure types a hexagonal net of cations results, and there are many variants of the stacking sequences resulting from the weak van der Waals interactions between the layers. Despite the common building blocks the MX_2 and MX_3 compounds have a wide variety of interesting magnetic structures and behaviors.

Structure and Luminescent Properties of Metal-organic Frameworks

Metal-organic frameworks (MOFs) are promising multifunctional luminescent materials because of their inherent advantages of containing both organic linkers and inorganic metal ions, as well as tailorability in terms of structure, dimension, size, and shape. Furthermore, the metal centers, organic linkers, metal-organic charge transfer, and guest molecules within porous MOFs all can potentially generate luminescence. Such uniqueness can allow us to generate luminescent MOF materials with systematically varied luminescence properties, which are crucial for the lighting, display, and optical devices. Although the development of luminescent MOFs is still in its infancy, the currently available results have unambiguously demonstrated that the design and construction of MOFs for luminescent functionality is an active field.

Barsukova et al. outlines four new Zn(II) and Cd(II)-based flexible MOF materials, with bis(imidazole-1-yl) alkane as the ligands: their synthesis, structural characterization, and solid-state luminescence measurements. The authors concluded that the structure of these compounds, particularly the length of the flexible linkers, has a great influence on the optical properties.

Semiconductor Porphyrins

Porphyrins are a group of heterocyclic macrocycle organic compounds, composed of four modified pyrrole subunits interconnected at their α carbon atoms via methine bridges (=CH−). The parent porphyrin is porphin, and substituted porphines are called porphyrins. The main functions of porphyrins, in addition to supporting aerobic life, include a variety of applications based on their optical, semiconducting, and redox properties. They can perform electronic functions in devices ranging from diodes to rectifiers, wires, and capacitors, and as materials in electronic or magnetic memory.

The crystallographic features of various porphyrins are fascinating and the various features govern the subsequent electronic properties. The structure of the material, particularly the planarity of the macrocycle and the conjugation of pi-bonding through the proximity of adjacent macrocycles or the interconnection of macrocycles via appropriate covalently-bonded ligands, play a pivotal role in the solid state semiconducting properties of porphyrins.

Cook et al. reviewed the structure of the parent porphyrin compounds, the peripheral substituents, the metalation of the macrocycle core, the role of covalent inter-molecular bonding, and the electronic properties as related to structure. Various factors, important in optimizing porphyrin semiconducting properties, were summarized and discussed at the molecular level, and in single crystals, thin films, and bulk polycrystalline materials. Finally, a number of structural examples of complex multiporphyrins were illustrated, including bridged multiporphyrins, confacial dimers, directly linked porphyrins, multiple-linked structures, self-assembled arrays, and metal-bridged arrays.

Photocatalysis Materials Constructed using α-U_3O_8 Types of Layers

The origins of a hydrogen energy economy go back to the 1920s, but it was not until the 1970s, an era where several energy/oil crises occurred, when the idea of using hydrogen to replace fossil fuels took traction and many scientists and engineers took up the challenges of producing and storing hydrogen. As the price of oil declined during the 1980s, so did the emphasis on finding alternatives. However, in the 2003 State of the Union Address President Bush stated, "A simple chemical reaction between hydrogen and oxygen generates energy, which can be used to power a car, producing only water, not exhaust fumes. With a new national commitment, our scientists and engineers will overcome obstacles to taking these cars from laboratory to showroom, so that the first car driven by a child born today could be powered by hydrogen, and pollution-free". With this statement, the hydrogen economy was back in focus, and research into the production and storage of hydrogen again briefly surged. In photoelectrochemical water splitting, the electricity used to produce the hydrogen is supplied by a renewable energy source (i.e., a photovoltaic system the series of mixed-metal oxide structures, based on the stacking of α-U_3O_8 type pentagonal bipyramid layers, with the general formula $A^{m+}{}_{((n+1)/m)}B_{(3n+1)}O_{(8n+3)}$, where the A cation can be Ag, Bi, Ca, Cu, Ce, Dy, Eu, Gd K, La, Nd, Pb, Pr, Sr, or Y, the B cation is either Nb or Ta, and m = 1–3 and n = 1, 1.5, 2. Some of the compounds from the series have phase transitions, associated with changes in temperature, composition, and chemical reactions, resulting in structural distortions that lower the symmetry of the crystal structure. Related to these structural distortions are electronic distortions that impact the efficiency of the photocatalysis process.

Cationic Doping in $[Ca_{24}Al_{28}O_{64}]^{4+}$ Clathrate Framework, an Electride

The compound $Ca_{12}Al_{14}O_{33}$ (Z=2) is located close to the middle of the CaO–Al_2O_3 binary phase diagram, a historic phase diagram used by many of the early ceramic industries and geologists, rich with intermediate compounds. $Ca_{12}Al_{14}O_{33}$, or mayenite, has long been identified with various cement mixtures. The crystal structure of $Ca_{12}Al_{14}O_{33}$ consists of a framework made up of the Ca and Al cations and 32 of the oxygen anions, forming a cage structure, with the remaining oxygen needed to charge balance the structure, occluded within the cage. Depending on the processing conditions, the occluded oxygen can be removed and replaced with other atoms, molecules, or electrons giving rise to a host of interesting properties including near metallic conduction. The ability for the electrons to remain in the room temperature has earned the compound the distinction of being the first room-temperature inorganic structure at electride. Many researchers have tried doping the stoichiometric oxygen occupied structure with various cations to improve the resulting properties, but few have continued to study the change in the physical properties and electron behavior in the electride. Cationic doped isostructural compounds and various processing methods with the purpose of invigorating research into the effects that dopants can have on the density of states, localized electron behavior, structural distortions, and ultimately the resulting electronic properties.

3

Types of Crystallography

There are various types of crystallography such as quantum crystallography, x-ray crystallography, electron crystallography, optical crystallography, isometric crystallography and tetragonal crystallography. Fourier space, Bragg's law of diffraction, etc. are some of the aspects that fall under its domain. All these different types of crystallography and related aspects have been carefully analyzed in this chapter.

Quantum Crystallography

Quantum crystallography is a branch of crystallography that investigates crystalline materials within the framework of quantum mechanics, with analysis and representation, in position or in momentum space, of quantities like wavefunction, electron charge and spin density, density matrices and all properties related to them (like electric potential, electric or magnetic moments, energy densities, electron localization function, one electron potential, etc.). Quantum crystallography involves both experimental and computational work. The theoretical part of quantum crystallography is based on quantum mechanical calculations of atomic/molecular/crystal wavefunctions, density matrices or density models, used to simulate the electronic structure of a crystalline material. Experimental works mainly rely on scattering techniques (X-rays, neutrons, γ-Rays, electrons), although spectroscopy as well as atomic microscopy are also sources of information.

The connection between crystallography and quantum chemistry has always been very tight, after X-ray diffraction techniques became available in crystallography. In fact, the scattering of radiation enables mapping the one-electron distribution or the elements of a density matrix. The kind of radiation and scattering determines the quantity which is represented (electron charge or spin) and the space in which it is represented (position or momentum space). Although the wavefunction is typically assumed not to be directly measurable, recent advances enable also to compute wavefunctions that are restrained to some experimentally measurable observable (like the scattering of a radiation).

Quantum chemistry methods and crystal structure determination are highly developed research tools, indispensable in today's organic, inorganic and physical chemistry. These tools are usually employed separately. Diffraction and scattering experiments provide structures at the atomic scale, while the techniques of quantum chemistry provide wavefunctions and properties derived from them.

The term quantum crystallography was first introduced by Massa, Huang and Karle in 1995 for methods that exploit "crystallographic information to enhance quantum mechanical calculations and the information derived from them". The basic idea is to compensate shortcomings and limitations of quantum mechanical models, e.g. incomplete consideration of electron correlation, with experimental data that are not suffering from the same limitations. In 1999 the same authors also suggested the converse possibility: "quantum mechanics can greatly enhance the information available from a crystallographic experiment".

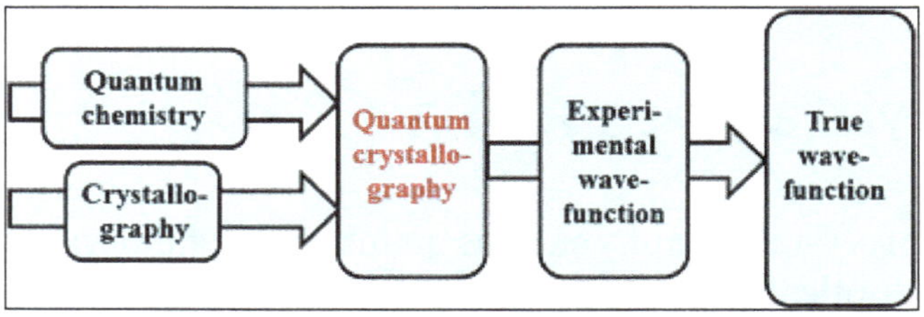

Quantum crystallography, first aspect: crystallographic data are integrated into quantum chemical calculations to enhance the information content of the wavefunction. The resulting, so-called "experimental wavefunction" represents an improved approximation to the true wavefunction.

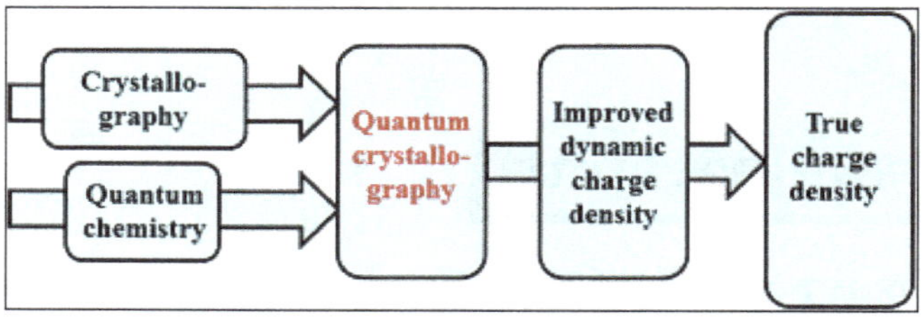

Quantum crystallography, second aspect: quantum chemical calculations are integrated into crystal structure determination to improve the dynamic charge density, i.e. the thermally smeared electron and nuclear densities.

Mutually subsidiary combinations of quantum chemistry and X-ray structure determination suggest themselves. Quantum chemical models are usually based on some approximations of the true wavefunction. X-ray structure determination aims at the true charge density, which is related to the square of the true wavefunction, but is affected to a smaller or larger extent by vibrational motion and experimental errors.

Here the topic of quantum crystallography is presented in two parts. The first part summarizes the development of increasingly sophisticated methods to combine information from quantum chemical calculations with diffraction and other experimental data. The second part describes ways of improving structural models obtained from diffraction experiments by combining them in a self-consistent way with information from quantum chemical calculations.

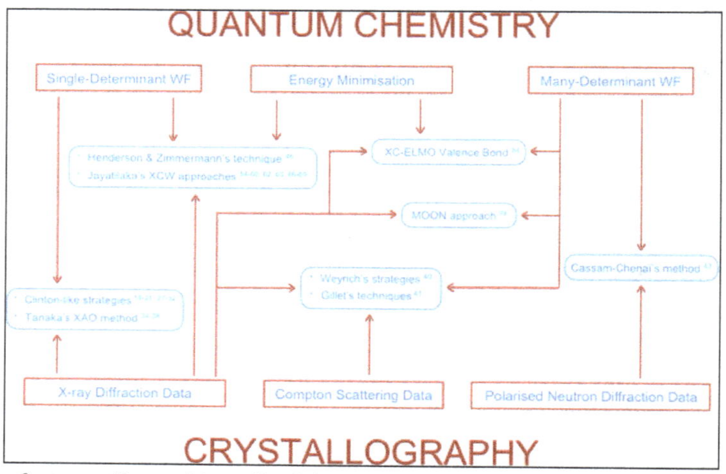

Scheme summarizing the features (framed in red) of the main methods (framed in light blue). The lower the position of the method is in the scheme, the lower is the quantum chemistry contribution in it.

Enhancing Quantum Chemical Calculations with Experimental Information

1. Pioneering experimental wavefunction techniques:

All modern quantum crystallography techniques according to the first definition originate from the pioneering "experimental" wavefunction strategies originally proposed in the 1960s by Mukherji and Karplus. They perturbed unconstrained Hartree–Fock molecular orbitals until they produced satisfactory agreement with experimental dipole moments or electric field gradients at a minimal increase in the energy of the system. Agreement improved not only for the constraining experimental values, but also for other properties, such as diamagnetic and paramagnetic susceptibilities. Rasiel and Whitman introduced experimental dipole moment constraints into the variational minimization of the energy with Lagrange multipliers. Byers Brown and Chong proposed to implicitly introduce the experimental constraints by adding proper quantum mechanical operators Ôi (multiplied by a Lagrange multiplier λi) to the starting and unconstrained Hamiltonian Ĥ:

$$\hat{H}' = \hat{H} + \sum_i \lambda_i \hat{O}_i$$

The modified Schrödinger equation can be solved variationally as shown by Fraga and Birss.

In 1969 Clinton et al. published a series of groundbreaking ideas that may be considered the direct precursors of today's quantum crystallography methods. Although X-ray diffraction data were not considered explicitly, the proposed theoretical framework forms the basis for the methods developed in the 1970s and 1980s that combine quantum mechanical calculations with crystallographic data. Here, the authors presented a semi-empirical strategy to determine one-electron density matrices of small diatomic molecules. They had both theoretical and practical reasons for focusing on one-electron density matrices rather than the more usual wavefunctions. To accomplish their task, they used theoretical constraints, such as the Hellmann–Feynman and virial theorems in combination with experimental constraints, e.g. potential energy curves obtained from spectroscopic and scattering measurements. However, the density matrices initially obtained by Clinton et al. did not conform with the essential quantum mechanical requirement that each density matrix must correspond to an antisymmetric wavefunction (N-representability). For single Slater determinant wavefunctions, this drawback can be overcome by imposing the idempotency condition (i.e., $P^2 = P$) and the normalization to the number of electrons (TrP = N) on the one-electron density matrices. This goal was achieved by combining the original, iterative McWeeny density matrix purification formula:

$$P_{n+1} = 3P_n^2 - 2P_n^3$$

with (i) the Hellmann–Feynman and virial theorems, (ii) Parr's integrated Hellmann–Feynman theorem, (iii) local energy constraints and, later, (iv) cusp conditions. This gave rise to another iterative procedure in which the density matrix at each iteration is obtained according to the following equation:

$$P_{n+1} = 3P_n^2 - 2P_n^3 + \sum_k \lambda_k{}^{(n)} O_k$$

where P_{n+1} and P_n represent the one-electron density matrix in the adopted finite set of basis functions at the $n + 1$ and n iterations; O_k is the matrix representation of the operator $\hat{O}_k$ associated with the property k in the same finite basis set as the density matrix. The external multipliers $\{\lambda_k^{(n)}\}$ are determined in a separate calculation by solving an auxiliary set of linear inhomogeneous equations that constrain the density matrix to reproduce the values $\{o_k\}$ of the properties considered.

2. The first quantum crystallographic methods: the original Clinton and Massa approach and its later developments-

An X-ray diffraction experiment provides a large number of structure factor amplitudes, observables from which the electron density in the crystal can be determined (after recovering their phases). Recognizing this fact Clinton and Massa attempted for the first time to extract a one-electron density matrix from X-ray diffraction data. Their 1972 technique may be considered as the first quantum crystallography investigation. It differs from the strategy proposed by Clinton and coworkers in 1969 only by using structure factors as external constraints and basically consists in iteratively solving the following matrix equation:

$$P_{n+1} = 3P_n^2 - 2P_n^3 + \lambda_N^{(n)} I + \sum_h \lambda_h^{(n)} f_h$$

Where, P_{n+1} and P_n have the same meaning as in $P_{n+1} = 3P_n^2 - 2P_n^3 + \sum_k \lambda_k^{(n)} O_k$, I is the identity matrix, f_h is the matrix of the Fourier transforms of the basis functions products, h is a triad of Miller indices labeling the reflection considered, $\lambda_N^{(n)}$ and $\lambda_h^{(n)}$ are external multipliers determined at each iteration from the auxiliary set of equations that constrain the density matrix to be normalized to the correct number of electrons and to reproduce the given set of structure factors.

Initially Clinton, Massa and others studied very small model systems at rest (mainly atoms and the hydrogen molecule) with structure factors calculated from near-exact wavefunctions. In 1973 they generalized their algorithm to take into account real experimental data with their errors (but did not test it with actual data). Because of the experimental errors, the structure factor constraints cannot be satisfied exactly but only in a least-square sense. Frishberg compared in detail the values of some physical properties for the beryllium atom predicted by the near Hartree–Fock wavefunction of Clementi on one hand and by the X-ray fitted and Best Density Matrix (BDM) wavefunctions on the other hand. In most cases the one-electron properties obtained with the quantum crystallography strategy agreed best with the results calculated from the highly accurate and correlated wavefunction used as reference. As far as momentum-related properties are concerned, the predictions based on the X-ray fitted wavefunction were always superior to the near Hartree–Fock ones. However, the BDM technique made the best predictions.

The investigations summarized so far established the theoretical basis for combining quantum chemical calculations with experimental X-ray diffraction data as external constraints, showed its potential and its limitations, but used only calculated X-ray structure factors (usually for atoms at rest). The first density matrix determined from experimental X-ray structure factors was that of beryllium in 1985. In a later study by Aleksandrov et al. one-electron density matrices for silicon and diamond were successfully fitted to experimental structure factors. Electronic kinetic energies and Compton profiles derived from them were in good agreement with corresponding experiments.

Over the years several modifications and improvements of the Clinton & Massa "experimental" density matrix method were proposed. Massa and coworkers extended the original approach (i) to the case of open-shell systems and (ii) to extended systems using Bloch and Wannier functions. The latter extension takes into account the interaction among unit cells and allows to study insulators, semiconductors and metals, at least in principle. The most important step forward for this family of techniques is probably the one proposed by Pecora. He devised a simplified steepest-descent algorithm to determine "experimental" idempotent density matrices. It is mainly based on the gradient of the average squared error between calculated and experimental structure factors amplitudes (*i.e.*, the well-known χ^2 statistic). With his approach, he provided one of the earliest density matrices obtained from real experimental data, positron annihilation data for a Cu–Ge alloy in his case. Pecora's algorithm was further improved by Howard *et al.* The resulting powerful simulated annealing strategy allowed for the first time to study molecules as large as methylamine and formamide, for which theoretical and experimental X-ray diffraction data were used as external constraints, respectively. In a follow-up investigation Snyder and Stevens determined the density matrix for the azide ion in potassium azide from experimental structure factors. They found that the electron density associated with the fitted density matrix shows some features that are common both to the multipole model and to theoretical charge distributions used for comparison.

Among the pioneering quantum crystallography methods, the strategy proposed by Figgis and collaborators has to be mentioned. They extracted populations of valence atomic orbitals from X-ray or polarized neutron diffraction data, *e.g.* in coordination complexes such as the *trans*-tetraamminedinitronickel(II) $[Ni(ND_3)_4(NO_2)_2]$ compound and the $CoCl_4^{2-}$ anion in Cs_3CoCl_5. The X-ray Atomic Orbital (XAO) method developed by Tanaka and coworkers also belongs to this family of approaches. It determines atomic orbitals on transition metal ions by minimizing the usual statistical agreement between experimental and theoretical structure factors subject to orbital orthonormality and symmetry constraints. Although the strategy initially aimed at modeling the crystal-field influence on the d atomic orbitals, it was later extended to also treat s, p and f functions.

3. Beyond the single Slater determinant approximation:

All the methods described so far suffer from an intrinsic limitation: since they require the one-electron density matrices to be idempotent, these matrices are necessarily associated with single Slater determinant wavefunctions. The MOON (Molecular Orbitals Occupation Numbers) method tries to overcome this problem by re-determining the occupation numbers of pre-computed occupied and virtual molecular orbitals, such that the agreement with the X-ray diffraction data is maximized. The current version of the method offers two options. In the first the optimized occupation numbers of the molecular orbitals are constrained to be between 0 and 2, which implies that one-electron density matrices are N-representable. In the other option, the upper limit on the occupation numbers is dropped. This introduces more flexibility in the fitting process, but may lead to results violating the Pauli principle. In both options the coefficients of the pre-determined molecular orbitals are not optimized during the fitting procedure; the technique thus scales linearly with the size of the basis set. However, the necessity to pre-compute the molecular orbitals renders the application of MOON to very large molecules quite difficult.

More advanced methods to overcome the single-determinant limitation replace the basic idempotency constraint with a more general N-representability condition that requires the eigenvalues for the one-electron density matrix to lie between 0 and 1. Weyrich and collaborators introduced this condition into their pioneering methods for the reconstruction of the complete one-electron density matrix (i.e. including diagonal and non-diagonal parts) from a combination of X-ray diffraction and inelastic Compton scattering data. Gillet and coworkers followed the same philosophy in more recent studies. Weyrich's technique was extended by Nicholson et al. to obtain Dyson orbitals from electron momentum spectroscopic data. The strategy devised by Cassam-Chenaï is included here in the same group of techniques; a small number of wavefunctions (obtained through preliminary *ab* initio calculations) are used to expand the ensemble N-representable one-electron density matrices of the systems investigated. The expansion coefficients are determined through a fit to the experimental measurements. Using this idea Cassam-Chenaï et al. determined the spin density of the $CoCl_4^{2-}$ anion in the Cs_3CoCl_5 crystal from polarized neutron diffraction data.

4. Criteria to select the best wavefunction:

The theoretical developments and practical strategies outlined above represent an important basis for combining quantum mechanical calculations with experimental data. However, as Gilbert has pointed out, they suffer from a further ambiguity. There is an infinite number of N-representable one-electron density matrices (differing only in their off-diagonal parts) and, consequently, an infinite number of wavefunctions that are compatible with a given electron distribution. Therefore, even under strict N-representability constraints, simple fitting to density data may lead to an inappropriate density matrix or wavefunction. Henderson & Zimmermann proposed that, among all single Slater determinants that reproduce a given electron density, the best one is the one that minimizes the Hartree–Fock energy functional. They introduced this criterion into the 1972 approach of Clinton & Massa. Their algorithm modifies in a quasi-continuous way a starting Hartree–Fock wavefunction until the desired X-ray constrained solution is obtained at a minimum penalty in energy. A similar philosophy was followed by Kryachko and coworkers. Their original method of local-scaling transformations was exploited to modify an initial wavefunction (the promolecule wavefunction) until it reproduces the electron density given by the experimental X-ray diffraction data. In this approach the energetic criterion of Henderson & Zimmermann was not considered explicitly.

Levy & Goldstein and later Gritsenko & Zhidomirov proposed an alternative to the Henderson & Zimmermann criterion. They select the single Slater determinant with minimal kinetic energy among the ones compatible with a given charge distribution. Zhao, Morrison and Parr used this idea in the early 1990s to extract Kohn–Sham orbitals from electron density data generated by *ab initio* computations. Orbitals obtained with this philosophy are almost identical with the corresponding Hartree–Fock ones. For this reason the authors proposed (i) that the Kohn–Sham single Slater determinants obtained by kinetic-energy minimization could represent physically meaningful wavefunctions and (ii) that their strategies could provide an alternative way towards proper wavefunctions from experimental data. Other techniques to extract wavefunctions and crucial DFT quantities from the electron density have previously been introduced by Nyden & Parr and by March and coworkers. Unfortunately, their strategies have never been tested with real experimental data as external constraints.

5. Jayatilaka's X-ray constrained wavefunction fitting: the original strategy and recent advances.

Nowadays, the most reliable and practical implementation of the Henderson & Zimmerman idea is the X-ray constrained wavefunction (XCW) fitting approach devised by Jayatilaka. The method is currently implemented for molecular crystals. It assumes identical, space group-symmetry related, non-interacting molecules. The unit-cell electron density of the molecular crystal is expressed in terms of the electronic wavefunction ψ of a reference molecule and its symmetry related copies. To guarantee that the unit-cell electron density of the fictitious non-interacting molecular crystal is identical to the unit-cell electron distribution of the corresponding real interacting system, the Jayatilaka method determines the wavefunction ψ of the reference molecular unit that minimizes a new functional given by the sum of the energy of the molecule and of an additional term measuring the statistical agreement between the experimental and the calculated structure factor amplitudes:

$$J[\psi]=\langle\psi|\hat{H}|\psi\rangle+\lambda\left(\chi^2[\psi]-\Delta\right)$$

Where, λ is an external multiplier manually adjusted during the "fitting procedure" until achieving the desired statistical agreement Δ between the theoretical and the experimental structure factors (typically fixed to 1.0). χ^2 is the statistical agreement between the calculated and the experimental structure factor amplitudes:

$$\chi^2[\psi]=\frac{1}{N_r-N_p}\sum_h\frac{\left(\eta|F_{calc}(h;\psi)|-|F_{obs}(h)|\right)^2}{\sigma_h^2}$$

With N_r the number of measured X-ray diffraction data, N_p the number of adjustable parameters, h the triads of Miller indices labeling the reflections, σ_h the standard uncertainty of $|F_{obs}(h)|$, and η a scale factor properly determined to minimize χ^2.

From a theoretical point of view, the XCW method is fully supported by the Levy–Lieb theorem, according to which the exact wavefunction of a system not only provides its electron density, but also minimizes the sum of the kinetic and the electron–electron repulsion energies. For the same reason, the Henderson & Zimmerman criterion introduced in the previous subsection is to be preferred to the one based on the minimization of only the kinetic energy.

The XCW fitting strategy was originally developed in the framework of the restricted Hartree–Fock approach, but was later extended to the unrestricted case, to the relativistic scalar second-order Douglas–Kroll–Hess and Infinite-Order Two Component (IOTC) methods. It is also available in conjunction with the Kohn–Sham formalism. Recently the Jayatilaka method was coupled with a technique developed by Stoll *et al.* to determine Extremely Localized Molecular Orbitals (ELMOs). The resulting X-ray constrained ELMO method allows the extraction of X-ray constrained molecular orbitals strictly localized on small molecular fragments, such as atoms, bonds or functional groups.

6. Applications of the XCW fitting technique:

The XCW fitting method does not only provide reliable electron density distributions. From XCWs, physical properties of materials, such as non-linear optical properties, can be determined. These properties cannot be obtained from experimental electron densities since the crucial many-electron contributions would be missing. In 2013, Hickstein *et al.* exploited the Jayatilaka technique to compute dipole moments, molecular polarizabilities, hyper-polarizabilities and refractive indices for four compounds: coumarin (1-benzopropan-2-one), DED ({4-[bis(diethylamino)-methylium] phenyl}dicyanomethanide), MBADNP (3,5-dinitro-2-[1-phenyl-ethyl]aminopyridine) and ZTS (zinc [tris]thiourea sulphate), all of them displaying interesting non-linear optical properties. The authors claim that the properties calculated from XCWs agree with those calculated at the Hartree–Fock level with an average statistical deviation of 20% from each other, which is comparable to the statistical deviation between Hartree–Fock and experimental values for molecular polarizabilities.

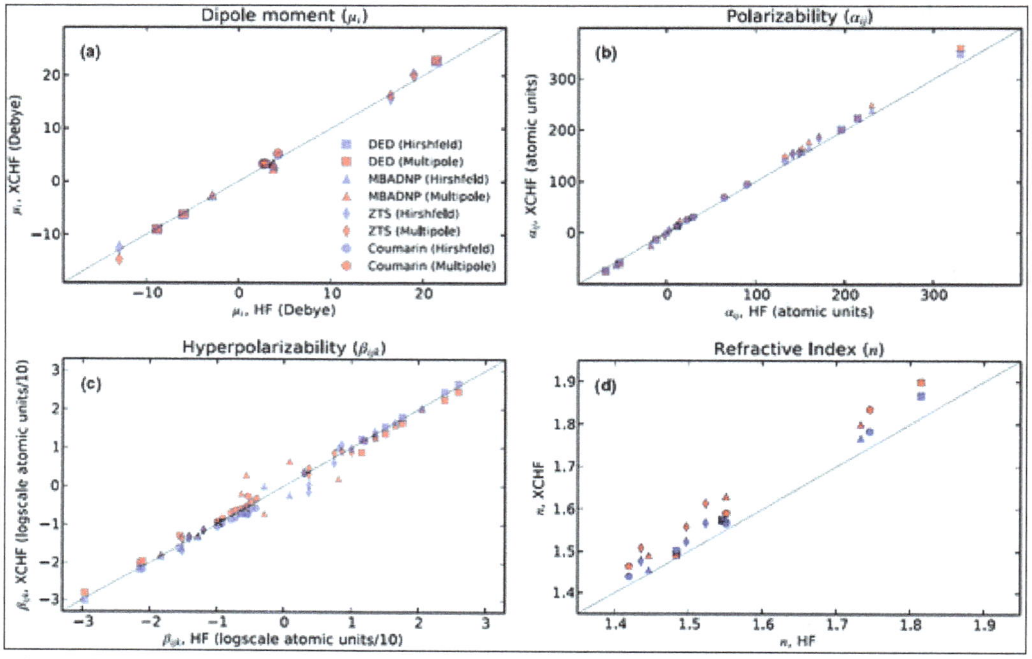

Comparison of optical properties calculated at the Hartree–Fock level with those calculated from XCWs. Two slightly different geometries and sets of atomic displacement parameters were used for the XCW fit, one based on a multipole refinement, the other on a Hirshfeld atom refinement model.

Another advantage of the XCW approach is its capability to include chemically important effects (*e.g.*, electron correlation and crystal effects) that can otherwise only be accounted for with a considerably larger computational effort. These can be studied by exploiting several functions and descriptors, such as the Electron Localization Function (ELF), the Electron Localizability Indicator (ELI) and bond indices, which were extracted from XCWs. The crystal structure of a vinyl sulfone (1-cyano-1-phenylsulfonyl-2-methylthio-2-methylaminoethylene) serves as an example. An ELI isosurface plot of the sulfonyl moiety is shown in figure; the most important Lewis resonance structures for the sulfonyl group are depicted in figure. In closed-shell systems the isosurfaces embrace regions of maximum localization of electron pairs and can hence be related to bonding and non-bonding valence effects. A topological analysis of the ELI provides basins for electron pairs with corresponding attractors. Each attractor can be interpreted as representing a bonding or lone electron pair. For the oxygen atoms there are two and three attractors and thus two and three lone pairs corresponding to one double and one single bond, respectively. This points towards an asymmetric Lewis description with an increased contribution of resonance form 2 (top). The asymmetry is induced by a strong intramolecular resonance assisted hydrogen bond of one of the oxygen atoms with a neighboring amine group. Interestingly, this asymmetry in the topology is not reproduced in a Hartree–Fock calculation where two lone pairs per oxygen atom are found, but is reproduced in a highly correlated CCSD (coupled-cluster with single and double excitations) calculation (unpublished results). It is known that ELI is particularly sensitive to electron correlation.

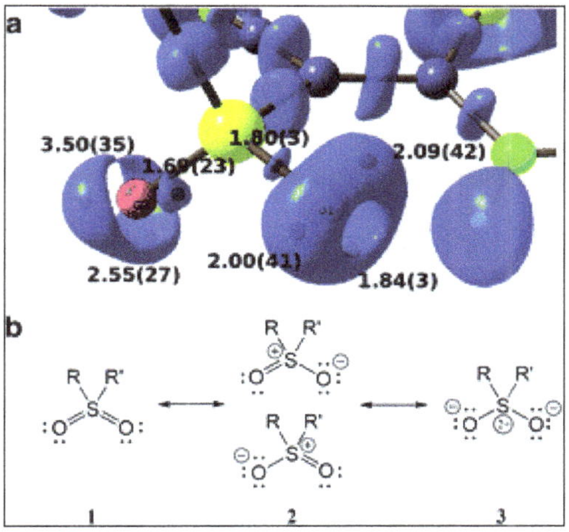

Figure shows (a) Isosurface representation of the X-ray constrained ELI for a sulfonyl group (ELI = 1.445 around the S–O bond attractors and ELI = 1.550 around the lone-pair attractors) with ELI valence basin electron populations (in e, with associated standard deviations). Black spheres below the transparent ELI isosurfaces indicate the positions of the basin attractors. Each basin attractor shown can be interpreted as representing a lone pair. The oxygen atom on the right is involved in an intramolecular hydrogen bond which induces asymmetry into the electronic system of the sulfonyl group.

7. Tampering with quantum chemistry:

From a quantum theoretical point of view, one might argue that the quantum crystallographic approaches tamper with clearly defined, albeit approximate, quantum mechanical models. The wavefunctions, one-electron density matrices and energies resulting from quantum crystallographic procedures are no longer solutions of the model Hamiltonian operator chosen for the quantum calculations, but are modified by the experimental data. There seems to be no rigorous mathematical proof based on the principles of quantum theory showing that this modification also represents an improved description of the system. The improvement can only be tested by comparing the modified description with the results of the best possible calculations or with experimental results not used as constraints. This has been done for optical properties calculated from XCWs. The XCW single Slater determinant produces better agreement with experimental data than the simple Hartree–Fock single Slater determinant. In order to test whether the Jayatilaka approach captures effects of electron correlation on the electron density, an XCW has been fitted to structure factors obtained from (highly correlated) coupled cluster computations. The comparison between the resulting electron distribution and the (theoretical) correlated one showed that electron correlation effects are partially captured if large values of the external multiplier λ are used and if mainly low-order reflections, which contain most of the information about electron correlation, are used as external constraints. These studies strongly indicate that XCW fitting strategies introduce information into the "experimental" wavefunction that is missing in the unconstrained wavefunction *ansatz*.

8. Towards enhancing models of nuclear motion with experimental information:

A comprehensive model of a crystal's charge density encompasses not only a description of the electron density, but also of nuclear motion. With a good quantum mechanical model and assuming the Born–Oppenheimer approximation it is possible to calculate not only the static equilibrium electron density, but also the crystal vibrations, *i.e.* the phonon dispersion spectrum of the crystal, which can be computed by diagonalizing the dynamical matrix. From the crystal vibrations, the mean square atomic displacement parameters (ADPs) are obtained. Such calculations can become extremely intricate depending on the density of k points used to sample the Brillouin zone and to model the $3nN$ vibrations for a crystal containing N unit cells with n atoms each.

This not only creates a need for simplified, yet accurate models, but also raises the question to what extent experimental mean square vibrational amplitudes (ADPs) can compensate for the simplifications. Hoser and Madsen have initiated what they call dynamic quantum crystallography. In this method, crystal normal modes and corresponding vibrational frequencies are determined from periodic *ab initio* calculations. The sampling of these modes is usually severely limited, often to the origin of the Brillouin zone (Γ-point). In the latter case the dynamical model encompasses only a single unit cell with n atoms and is thus highly oversimplified. The frequencies of the lowest crystal vibrations, the acoustic and low optical phonons, are particularly poorly represented in this approximation. These frequencies are therefore refined against experimental data until the ADPs derived from them, from the unmodified higher frequency phonons and from the refined atomic positions produce model structure factors F_{model} in agreement with X-ray or neutron diffraction data (Normal Mode Refinement, NoMoRe). Refined frequencies and calculated atomic displacement patterns compare well with results obtained independently from multi-temperature X-ray diffraction studies. Thermodynamic properties from NoMoRe and multi-temperature X-ray and neutron diffraction investigations agree – albeit not perfectly – with each other and with calorimetric data. NoMoRe against X-ray data produces hydrogen atom ADPs that compare well with those from neutron diffraction data.

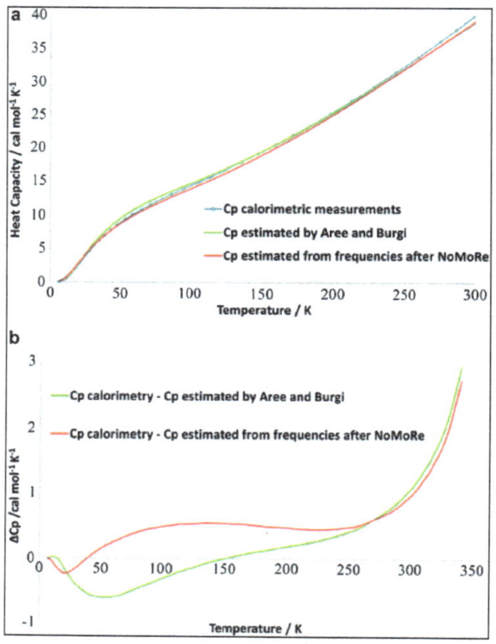

(a) Heat capacity (C_p) for a molecular crystal of naphthalene: from thermodynamic measurements (blue curve), calculated from frequencies obtained after NoMoRe (red curve) and estimated by Aree and Bürgi from multi-temperature diffraction data (green curve). (b) Difference between C_p

measured calorimetrically and C_p estimated with NoMoRe (red curve) or from multi-temperature data (green curve).

Enhancing Experimental Crystal Structure Models with Information from Quantum Chemical Calculations

1. Crystal structure determination by X-ray and electron diffraction has always used quantum-mechanical information:

It is often forgotten that the standard models of crystal structure determination incorporate the experimentally determined structure factor amplitudes on one hand and atomic form factors calculated independently by *ab initio* methods on the other. Such models depend to a variable extent on the calculated information introduced into them. To illustrate this point, it is recalled that the measured X-ray diffraction intensities $I_{obs}(h)$ with Miller indices $h = (h,k,l)$ are proportional to the square moduli of the structure factors $|F_{obs}(h)|$:

$$I_{obs}(h) \sim |F_{obs}(h)|^2$$

The structure factors F(h), which are the Fourier transforms of the crystal unit-cell electron density,

$$F(h) = \int dr\, \rho(r) e^{i2\pi h - r}$$

are modeled as a sum of atomic contributions:

$$F_{model}(h) = \sum_{j=1}^{N_{atoms}} f_j(h) T_j(h, U_j) e^{i2\pi h - R_j}$$

Where, f_j is the scattering or form factor of atom j at position R_j, and T_j is usually called 'atomic temperature factor', which depends on the mean square static and dynamic atomic displacements U_j of atom j from its average position R_j. U_j and R_j are variables of F_{model} and are refined against the experimental data by minimizing the mean-square difference between $|F_{obs}(h)|$ and $|F_{model}(h)|$. The atomic electron densities, *i.e.* the scattering factors f_j are not. From the beginning of X-ray crystal structure determination they have been taken as the Fourier transforms of calculated spherical electron densities of isolated atoms. In this model the f_j are assumed to always be the same, irrespective of an atom's chemical environment (Independent Atom Model, IAM). Early on, the atomic densities have been computed with the classical laws of electrodynamics, while later on they have been determined with increasingly sophisticated quantum mechanical models using, for example, Thomas–Fermi–Dirac or Hartree–Fock–Slater approximations. Most current software for crystal structure determination contains libraries with quantum-mechanical form factors calculated in the 1960s using relativistic wavefunctions. Reduced or increased scattering power of ions relative to the neutral atom is taken into consideration by Hartree–Fock calculations for lighter atoms, or relativistic Dirac–Slater calculations for heavier atoms.

In contrast to X-ray diffraction, conventional electron diffraction experiments measure the interaction of the probing electrons with the electrostatic potential inside the crystal. Intensities $I(h)$ are proportional to the square of the Fourier transform of the potential distribution $\varphi(r)$. The latter can be modelled with isolated-atom scattering factors similar to $F_{model}(h)=\sum_{j=1}^{N_{atoms}} f_j(h)T_j(h,U_j)e^{i2\pi h-R_j}$. Analytical Gaussian or polynomial fits to X-ray scattering factors are used to obtain numerical values of the electron-diffraction scattering factors. Note that the nuclear scattering lengths and cross sections needed in neutron diffraction are obtained experimentally.

2. Aspherical scattering factors and electron-density research:

The crystal field is never spherically symmetric, especially not if an atom forms strong directed bonds. Representing the crystal electron density as a sum of spherical atomic densities can thus be a severe approximation. Quantum-chemical calculations on molecules and polyatomic ions can account for distortions from spherical symmetry, but require procedures for partitioning the crystal electron density into a sum of atomic ones. The improved scattering factors are referred to as generalized or polarized-atom scattering factors. Stewart *et al.* suggested such scattering factors obtained from SCF calculations on diatomic and simple organic molecules. The treatment of the hydrogen atom is especially problematic, since hydrogen has no core electrons. The distribution of its single 1s-valence electron tends to be heavily deformed in covalent bonds. A scattering factor based on quantum-mechanical calculations of the H_2 molecule has been especially successful for the treatment of such hydrogen atoms. A spherical approximation of the scattering factor of a hydrogen atom in the hydrogen molecule is still in use in most modern crystallographic software (Stewart–Davidson–Simpson, SDS, model).

A different way to represent the crystal electron density in terms of aspherical atomic electron densities and their scattering factors is to parameterize pseudo-atom electron densities with analytical functions. The forms of the latter are based on quantum-chemical considerations, whereas their parameters are refined against the observed structure factors. Hence, such scattering factors are partly based on quantum mechanical theory and partly on diffraction data. Models in this family are known as multipole models. An incomplete list of the approaches that have been introduced and tested over the past 50 years. The most widely used formalism is that of Hansen and Coppens from 1978. Its aspherical scattering factor expression is shown in equation below:

$$f_j(h)=P_{f,core}\,f_{f,core}(h)+P_{f,val}\,f_{f,val}(h,\kappa)+\sum_{l=0}^{4}\phi_{jl}(h,\kappa'_1)\sum_{m=-l}^{l}P_{jlm}\,Y_{lm}(h/h)$$

The core and valence spherical atomic scattering factors $f_{j,core}$ and $f_{j,val}$ are usually derived from Hartree–Fock wavefunctions expanded over Slater-type basis functions and tabulated in data-bases. Similarly the ϕ_{jl} functions are Fourier–Bessel transforms of radial density functions R_{jl} based on single-zeta Slater-type orbitals with energy-optimized exponents taken from valence-orbital wavefunction calculations. $Y_{lm}{}^{j}$ are spherical harmonics. This means that the analytical form of the electron density is significantly influenced by quantum-chemical concepts and results, whereas the multipole population parameters P_{jlm}, the expansion–contraction parameters κ and κ' for the radial parts of the electron densities, positional parameters and ADPs are refined against the experimental diffraction data. In an effort to increase the flexibility of the multipole model, bonded-atom radial functions have been introduced. In an extended version of the Hansen–Coppens formalism, aspherical $f_{j,core}$ account for core deformation and can be refined against the experimental data in exceptional cases.

Atomic multipole descriptions derived from experiment and pertaining to different chemical environments have been collected in libraries of pseudoatoms (ELMAM). Multipole and expansion–contraction parameters have also been extracted from quantum-chemically computed electron densities of model compounds and tabulated for many atoms in a large but limited number of different chemical environments (invariom database, UBDB). Using aspherical rather than spherical scattering factors in crystallographic refinements of positional parameters and ADPs improves the accuracy of the structural model.

3. Quantum crystallographic methods according to the second definition:

Crystal structure determination using the IAM or pseudo-atom models always involves quantum-mechanical information. Certain aspects of the crystal electron density are not measured, but modeled with the help of theory. From this viewpoint, every X-ray structure determined since the appearance of the Schrödinger equation in 1925 could be regarded as 'quantum crystallography'. However, spherical atoms or pseudo-atoms in a standard chemical environment never reflect the influence of the actual chemical environment and the crystal field exactly. The quantum-chemical information introduced into the conventional structure modeling and refinement process is always approximate. The use of quantum chemistry in these methods may be considered as "indirect", since it relies on tabulated information for model compounds.

In the spirit of Huang, Massa and Karle's definition of quantum crystallography in 1999, we consider quantum crystallography as an interdisciplinary approach that amalgamates numerical quantum chemical information into diffraction experiments within single, integrated tools. The use of the wavefunction is "direct" since it is an integral part of the refinement and modeling process, is tailor-made for the very molecular or extended system under investigation and is continuously updated during structure refinement.

Quantitative convergent-beam electron diffraction (CBED) depends on such a dynamic model as it does not rely on tabulated atomic scattering factors. Each experiment requires its individual wavefunction to extract scattering amplitudes or Fourier potentials. To describe the multiple scattering between the incoming electron beam and the crystalline material a dynamical theory must be adopted. Dynamical Bloch-wave theory is often applied, which has some computational advantages for subsequent pattern matching procedures, and is similar to Pendellösung approaches for X-ray diffraction. The Schrödinger equation is solved for the single high-energy electron as it passes through the crystal and the electrostatic potential matrix elements are computed. In structure refinements from quantitative CBED data, simulated intensities are compared with observed intensity variations within CBED disks of different incident beam orientation. During refinement, wavefunction parameters in the dynamical theoretical model have to be updated repeatedly by adjustment of structural parameters and Debye–Waller factors. This "direct" combination of a diffraction experiment with theory can hence be considered the first quantum crystallographic technique according to the second definition. Method development and early applications date back to the 1960s. First interpretations of CBED disks with quantum chemical calculations appeared in the mid-1980s and were followed in the early 1990s by improved methods that take advantage of supercomputer facilities.

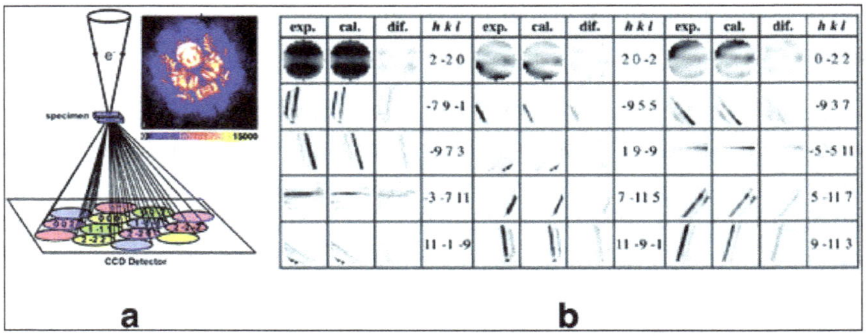

In figure, (a) Sketch of the scattering process leading to CBED disks that can be attributed to individual structure factors with Miller indices *hkl*. Intensity of the real measured pattern for aluminium in the crystal orientation 2 $\bar{2}$ 0 is shown as an inlay.(b) Final results of the fitting of the 2 $\bar{2}$ 20 Bragg pattern of silicon giving rise to several very accurate structure factors from which the electron density distribution could be modelled.

Quantitative CBED leads to very accurate electron density distributions, but only for simple materials with few independent atoms in the unit cell. They are much more accurate than those based on X-ray diffraction because (i) scattering factors for electrons at low scattering angles are acutely sensitive to bonding, (ii) dynamic scattering effects are modelled, and (iii) the amount of data in a single CBED pattern provides orders of magnitude more data points than unknowns. Nakashima *et al.* demonstrated in 2011 that the bonding electron-density distribution of aluminium from quantitative CBED is more reliable than that from X-ray diffraction or theory and that conclusions about anisotropic elastic constants can be drawn. Careful quantitative CBED experiments by Zuo *et al.* revealed bonding features between two copper atoms in Cu_2O. The deformation electron density around the copper atoms was interpreted in terms of a shift of electron density from a 3d orbital into a 4s orbital at the Cu atom. Static difference density maps (deformation density maps) between models based on aspherical and spherical atomic scattering factors are not unusual in X-ray diffraction. However, the accuracy of the result from quantitative CBED is remarkable. Palatinus *et al.* have recently shown that hydrogen atoms can be located with precession electron diffraction tomography on nanocrystals if dynamical refinement is used. This finding complements the analogous capabilities of Hirshfeld atom refinement.

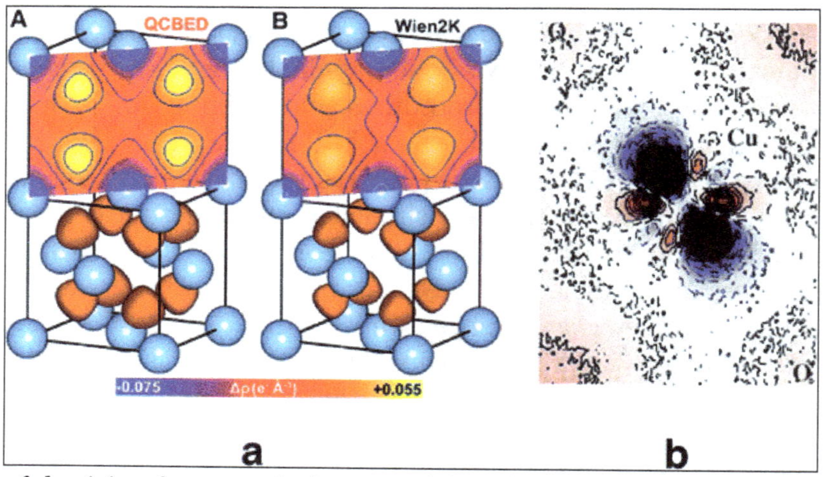

(a) Electron density of aluminium from quantitative CBED (left) and a periodic plane-wave calculation using the program Wien2k (right). (b) Static deformation density map around a copper atom in Cu_2O.

The first quantum crystallographic method in small-unit cell X-ray crystallography was introduced in 2008 by Jayatilaka and Dittrich. It is a two-step procedure called Hirshfeld Atom Refinement (HAR). In the first step, a molecular electron density is calculated from the atomic coordinates of an input geometry, typically at the Hartree–Fock or DFT level of approximation. The resulting electron density ρ_{mol} is partitioned into aspherical atomic electron densities ρ_j (Hirshfeld atoms) using the stockholder partitioning,

$$\rho_j(r) = w_j(r)\rho_{mol}(r)$$

with,

$$w_j(r) = \frac{\rho_j^o\left(r - \widetilde{R}_j\right)}{\sum_i \rho_j^o\left(r - \widetilde{R}_i\right)}$$

$w_j(r)$ is the weight function of atom j which carves an aspherical electron density out of the molecular electron density. It is defined as the ratio between the spherically averaged atomic electron density $\rho^0{}_j\left(r - \widetilde{R}_j\right)$ of atom j at position $\widetilde{R}_j \|$ and the sum of the spherical electron densities of all other atoms in the molecule. The spherically averaged atomic electron densities $\rho^0{}_j\left(r - \widetilde{R}_j\right)$ are not taken from tables but calculated with the same quantum chemical model and basis-set as used for ρ_{mol}. Optionally, atomic Hirshfeld charges and dipole moments can be obtained from the calculated molecular electron density and placed on atomic sites surrounding the central molecule in order to polarize the molecular electron density, thus simulating crystal field effects. In the field of these charges and dipoles a new molecular electron density and new charges and dipoles are calculated and the process iterated until self-consistency is reached. In preparation for the second step, namely the structural least-squares refinement, the static Hirshfeld atoms $\rho_j(r)$ are Fourier transformed to get the static, aspherical Hirshfeld scattering factors f_j. They are included together with the usual temperature factor T_j for each atom j in the structure factor $F_{model}(k)$:

$$F_{model}(k) = \sum_{j=1}^{N_{atoms}} f_j(k;\psi)T_j(k,U_j)e^{ik - \widetilde{R}_j}$$

Where, ψ stresses the dependence of the static Hirshfeld scattering factors on the calculated wavefunction. After refinement of the atomic positions $\widetilde{R}_j$ and the atomic displacement parameters U_j, a new electron density is calculated with the updated atomic coordinates. In an automated and iterative version from 2014, the two-step process is repeated until the parameter shifts in $\widetilde{R}_j$ and U_j divided by their standard uncertainty are less than 0.01. A somewhat simplified version of HAR has recently become available in the standard Olex2 crystallographic software.

The HAR procedure has advantages and disadvantages. On the one hand it ensures tailor-made aspherical atomic scattering factors. They optimally represent the unit-cell electron density within the limits of the chosen quantum mechanical procedure. This leads to enhanced crystal structure information in the spirit of the second definition of quantum crystallography. On the other hand the computational cost (CPU time) for a HAR refinement is significantly higher than for an IAM or multipole model refinement. Since the original HAR procedure is based on a molecular wavefunction, the treatment of periodic network structures is difficult. A recent variant of HAR based on periodic wavefunctions renders possible the treatment of network structures, and is potentially more accurate, but even more time-consuming.

First results obtained with HAR procedures are encouraging in two ways. Firstly, as Wall points out: "the results indicate that HAR can yield not only molecular geometries and ADPs that are similar to the neutron crystal structure, but also both 2Fo–Fc maps and static charge densities that are distinct from the multipole model, but that nevertheless agree comparably with the experimental data. Quantum crystallography therefore can yield accurate charge densities that are consistent simultaneously with theory and experiment". Chęcińska *et al.* and Dittrich *et al.* have reached similar conclusions in earlier electron-density research involving HAR. Secondly, several groups have shown that for data sets of good quality, hydrogen atomic positions and ADPs can be freely refined and are generally found in quantitative agreement with results from neutron diffraction. This holds for a large range of organic molecules and for data sets with resolutions as low as 0.8 Å, which can routinely be measured at in-house diffractometers with conventional X-radiation sources.

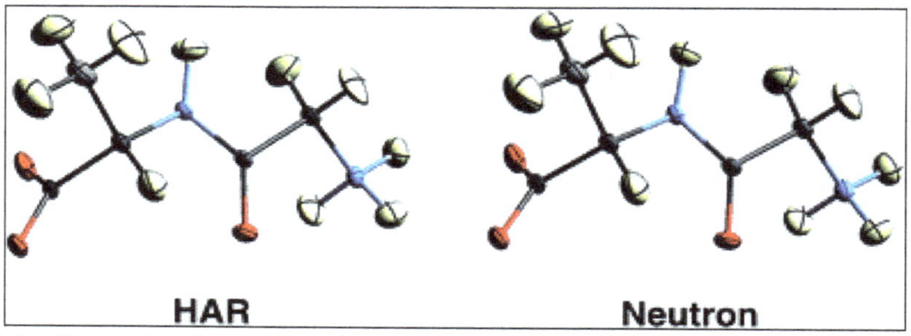

150 K structures of the dipeptide Gly-L-Ala with anisotropic displacement parameters (ADPs) at 50% probability from HAR based on X-ray diffraction data (left) and from neutron diffraction data (right). The structures are virtually identical by visual inspection.

Table: Mean absolute differences and their population standard deviations for the dipeptide Gly-L-Ala in terms of bond distances d involving hydrogen atoms and ADPs U_{ij} of hydrogen atoms between HAR (based on X-ray diffraction) and neutron diffraction results for four different temperatures.

T in K	$\langle\lvert\Delta d\rvert\rangle$ in Å	$\langle\lvert\Delta U_{ij}\rvert\rangle$ in Å²
12	0.0085(51)	0.0047(42)
50	0.0093(52)	0.0040(34)
150	0.0087(62)	0.0037(31)
295	0.0125(84)	0.0075(58)

4. Quantum crystallography in protein structure determination:

Macromolecular crystallographic refinement is another area in which quantum mechanical calculations can supplement the information provided by X-ray diffraction measurements. As is well known, one of the main obstacles preventing atomic-level resolutions of macromolecular crystallographic structures is the limited resolution and number of diffraction data compared to the number of parameters needed to model atomic positions and thermal motion. To overcome this drawback, constraints and restraints are usually introduced in the refinement. While the former reduce the number of structural parameters to be determined, the latter introduce stereochemical information into the analysis that complements the one provided by the X-ray data. Restraints take the form of penalty terms dependent on the deviation from reference parameters – similar to energy penalties in molecular-mechanics (MM) force fields, although with some differences: (i) the reference parameters are generally obtained from a statistical analysis of a large number of accurate protein and polynucleotide crystallographic structures, as done by Engh & Huber who exploited the Cambridge Structural Database (CSD); (ii) electrostatic interactions and van der Waals attractions are neglected. With such tailor-made restraints, accurate structures of proteins have been determined. However, when a macromolecule harbours a small molecule, *e.g.* a ligand, an inhibitor, or a cofactor with or without a metal atom, accurate restraint parameters for the molecule are often not available and an accurate structural model is difficult to achieve.

To alleviate this problem, Ryde and coworkers developed the ComQum-X method. It combines information from multi-scale Quantum Mechanical/Molecular Mechanical (QM/MM) calculations with that from experimental structure factors. The following functional is minimized:

$$E_{ComQum-X} = E_{QM/MM} + \omega_{X-ray} E_{X-ray}$$

$E_{QM/MM}$ is the QM/MM energy of the investigated system; $E_{X\text{-ray}}$ is the crystallographic penalty function, which measures the agreement between the observed and the calculated structure factors; $\omega_{X\text{-ray}}$ is a weight that balances the computational and the experimental information. The QM part encompasses the active site of proteins, their ligands and the nearby residues. The rest of the system is treated at the molecular mechanics level. The QM/MM calculation for the entire system is part of each step of the refinement, in agreement with the quantum crystallography idea. Crystallographic structures of several proteins containing ligands have been refined with the ComQum-X technique.

A similar QM/MM approach was implemented and applied by Merz and coworkers. They successfully assigned protonation states of key residues in β-secretase, crucial in the pathogenesis of Alzheimer's disease and reliably described the zinc coordination in zinc metalloenzymes.

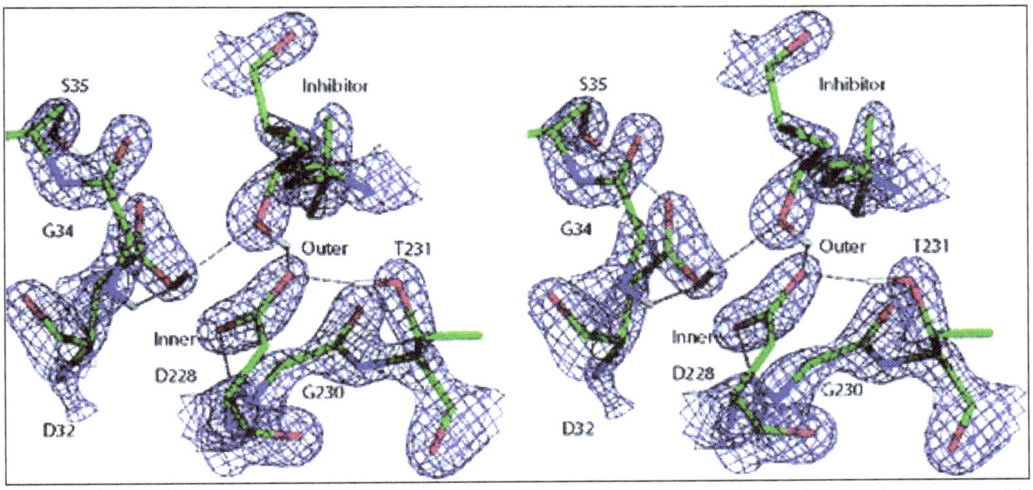

Cross-eye stereo view of the $2F_o - F_c$ electron density map (2.7σ contour level) around the active site residues of the β-secretase enzyme after QM/MM refinement of the most favoured 32i state. Note the hydrogen bond formed between the hydroxyl group of the inhibitor and the carboxyl group of residue D32.

The Merz group also proposed to replace the QM/MM computations of the ComQum-X approach by semiempirical quantum mechanical calculations. In order to be able to treat large molecular systems, they exploited their linear scaling "Divide & Conquer" (D&C) semiempirical method implemented in the program DivCon. The functional to be minimized now reads:

$$E_{DivCon-X} = E_{D\&C} + \omega_{X-ray} E_{X-ray}$$

$E_{D\&C}$ is the semiempirical quantum mechanical energy of the system from the "Divide & Conquer" approach. As in the QM/MM example, the D&C calculations are performed at each step of the refinement.

The method was tested by refining the crystallographic structure of the bovine pancreatic trypsin inhibitor. The QM refined structure and the one resulting from an Engh & Huber (EH) refinement were compared to the structure obtained from a conventional joint refinement of X-ray and neutron diffraction data not applying any restraints (Protein Databank (PDB) code 5PTI). The "Divide & Conquer" strategy provided improvements, mainly for hydrogen atom coordinates compared to the approach based on the EH restraints. The main reason for this improvement is the direct inclusion in the fully quantum mechanical refinement of electrostatics which is completely missing in the EH-based technique. The D&C strategy is now included in the popular crystallographic package PHENIX. The new, slightly different, algorithm was tested against a set of fifty protein-ligand structures obtained from the PDB. It always provided completely plausible ligand geometries and low ligand strains, also in cases in which the original, traditional refinement showed some difficulties.

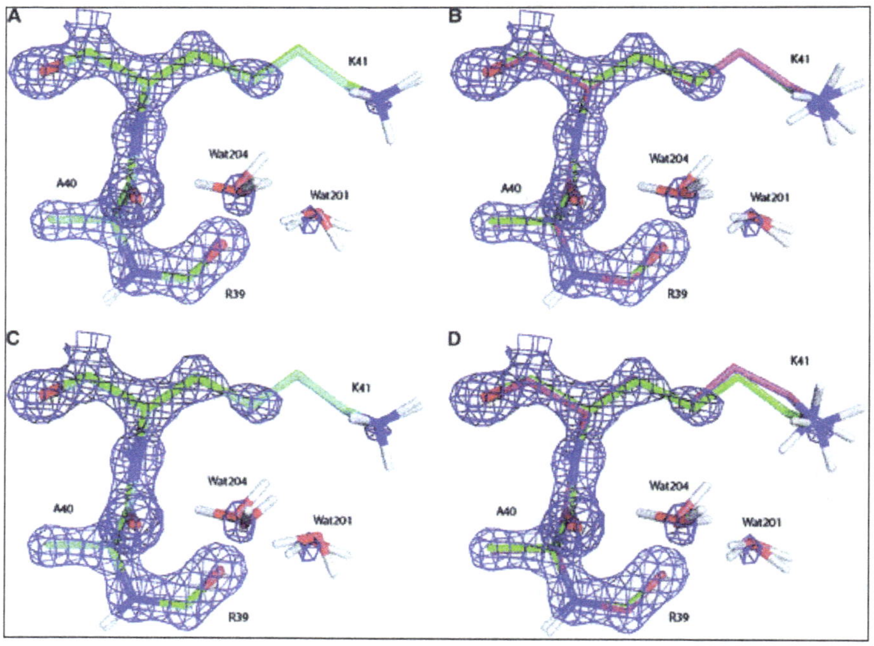

Electron density maps of residues Ala40, Lys41 and Arg30 (and two water molecules) of the bovine pancreatic trypsin inhibitor (2.3σ contour level) after (A) a full QM refinement with $\omega_{X\text{-ray}}$ = 0.9 (cyan), (B) an EH-restrained refinement with $\omega_{X\text{-ray}}$ = 0.9 (magenta), (C) a full QM refinement with $\omega_{X\text{-ray}}$ = 0.2 (cyan), (D) an EH-restrained refinement with $\omega_{X\text{-ray}}$ = 0.2 (magenta). All the obtained structures are superimposed to the 5PTI structure (green). Note the differences in hydrogen atom positions (aliphatic hydrogen atoms omitted for clarity).

X-ray Crystallography

X-ray crystallography is a technique used for determining the high-resolution, three-dimensional crystal structures of atom and molecules and has been fundamental in the development of many scientific fields. In its first decades of application, it is mainly used for determining the size of atoms, the lengths and types of chemical bonds, the atomic-scale differences among various materials, as well as the crystalline integrity, grain orientation, grain size, film thickness and interface roughness of the related materials, especially minerals and alloys. This method could also reveal the structure and function of many biological molecules like vitamins, drugs, proteins and nucleic acids. Up to date, it is still the chief method for characterizing the atomic structure of new materials and in discerning materials that appear similar by other experiments.

X-ray crystal structures can also explain the unusual electronic or elastic properties of a material, shed light on chemical interactions and processes, or function as the basis for designing pharmaceuticals against diseases. In particular, protein have been extensively put into structure determination by X-ray crystallography, which is also employed routinely in determining how a pharmaceutical drug interacts with its protein target and what changes might improve it. X-ray crystallography method has advantages of no damage to samples, free of pollution, low environmental requirements, high performance and precision over other measuring tools. These advantages make X-ray crystallography the most convenient and important manner to investigate the microstructure of materials.

X-rays are electromagnetic radiation with wavelengths between about 0.02 Å and 100 Å (1Å = 10^{-10} meters). They are part of the electromagnetic spectrum that includes wavelengths of electromagnetic radiation called visible light which our eyes are sensitive to (different wavelengths of visible light appear to us as different colors). Because X-rays have wavelengths similar to the size of atoms, they are useful to explore within crystals.

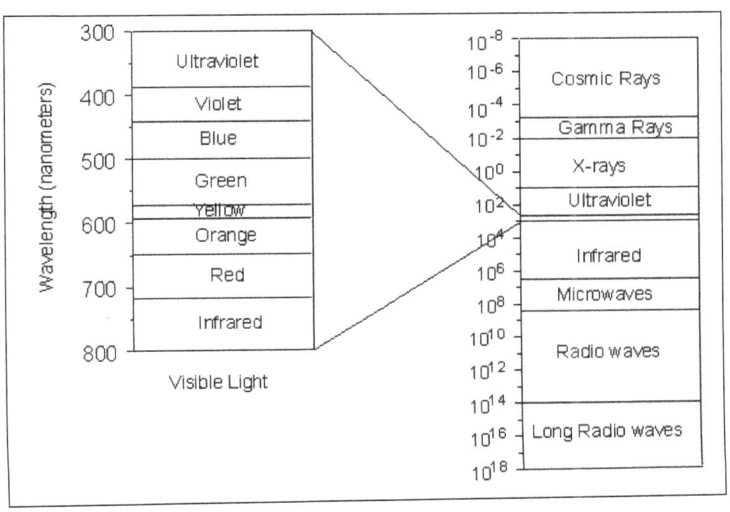

The energy of X-rays, like all electromagnetic radiation, is inversely proportional to their wavelength as given by the Einstein equation:

$$E = h\nu = hc / \lambda$$

Where,

- E = energy.
- h = Planck's constant, 6.62517 x 10-27 erg. sec.
- v = frequency.
- c = velocity of light = 2.99793 x 1010 cm/sec.
- λ = wavelength.

Thus, since X-rays have a smaller wavelength than visible light, they have higher energy. With their higher energy, X-rays can penetrate matter more easily than can visible light. Their ability to penetrate matter depends on the density of the matter, and thus X-rays provide a powerful tool in medicine for mapping internal structures of the human body (bones have higher density than tissue, and thus are harder for X-rays to penetrate, fractures in bones have a different density than the bone, thus fractures can be seen in X-ray pictures).

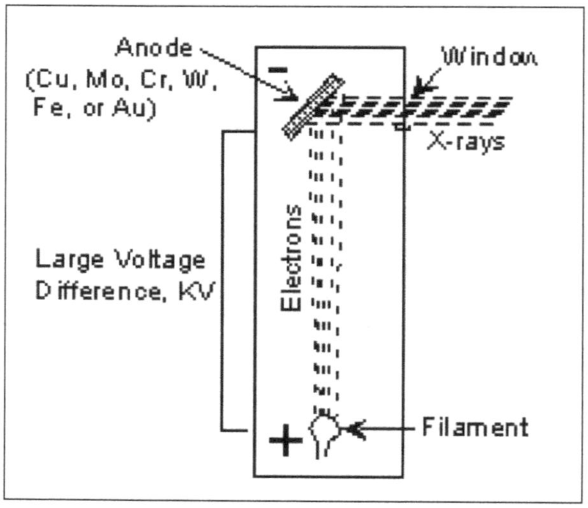

X-rays are produced in a device called an X-ray tube. Such a tube is illustrated here. It consists of an evacuated chamber with a tungsten filament at one end of the tube, called the cathode, and a metal target at the other end, called an anode. Electrical current is run through the tungsten filament, causing it to glow and emit electrons. A large voltage difference (measured in kilovolts) is placed between the cathode and the anode, causing the electrons to move at high velocity from the filament to the anode target. Upon striking the atoms in the target, the electrons dislodge inner shell electrons resulting in outer shell electrons having to jump to a lower energy shell to replace the dislodged electrons. These electronic transitions results in the generation of X-rays. The Xrays then move through a window in the X-ray tube and can be used to provide information on the internal arrangement of atoms in crystals or the structure of internal body parts.

Continuous and Characteristic X-ray Spectra

When the target material of the X-ray tube is bombarded with electrons accelerated from the cathode filament, two types of X-ray spectra are produced. The first is called the continuous spectra.

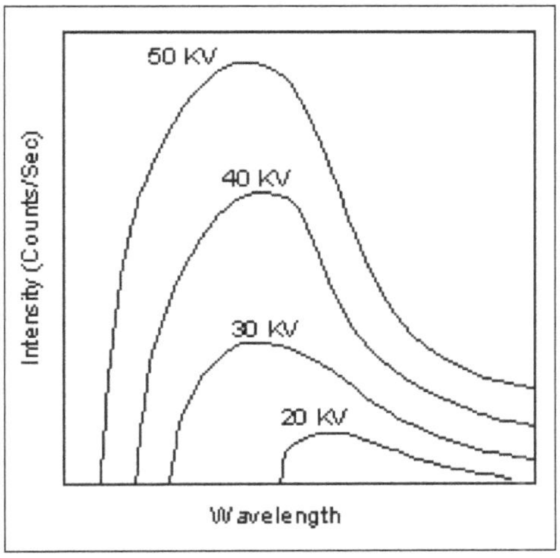

The continuous spectra consist of a range of wavelengths of X-rays with minimum wavelength and intensity (measured in counts per second) dependent on the target material and the voltage across the X-ray tube. The minimum wavelength decreases and the intensity increases as voltage increases.

The second type of spectra, called the characteristic spectra, is produced at high voltage as a result of specific electronic transitions that take place within individual atoms of the target material.

This is easiest to see using the simple Bohr model of the atom. In such a model, the nucleus of the atom containing the protons and neutrons is surrounded by shells of electrons. The innermost shell, called the K- shell, is surrounded by the L- and M - shells. When the energy of the electrons accelerated toward the target becomes high enough to dislodge K- shell electrons, electrons from the L - and M - shells move in to take the place of those dislodged.

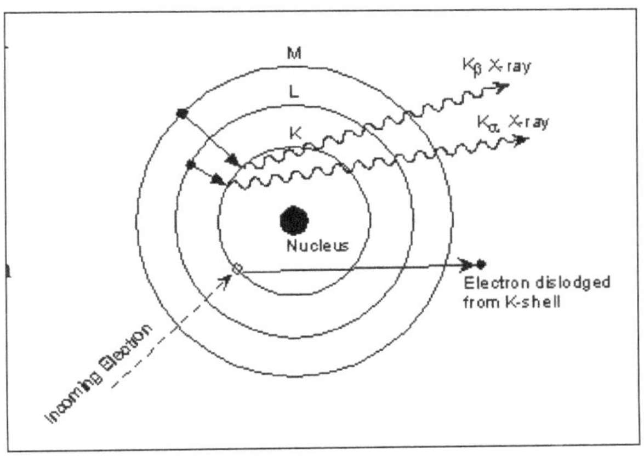

Each of these electronic transitions produces an X-ray with a wavelength that depends on the exact structure of the atom being bombarded. A transition from the L - shells to the K- shell produces a K_α X-ray, while the transition from an M - shell to the K- shell produces a K_β Xray.

These characteristic X-rays have a much higher intensity than those produced by the continuous sprectra, with K_α X-rays having higher intensity than K_β X-rays. The important point here is that the wavelength of these characteristic x-rays is different for each atom in the periodic table (of course only those elements with higher atomic number have L- and M - shell electrons that can undergo transitions to produce Xrays). A filter is generally used to filter out the lower intensity K_β X-rays.

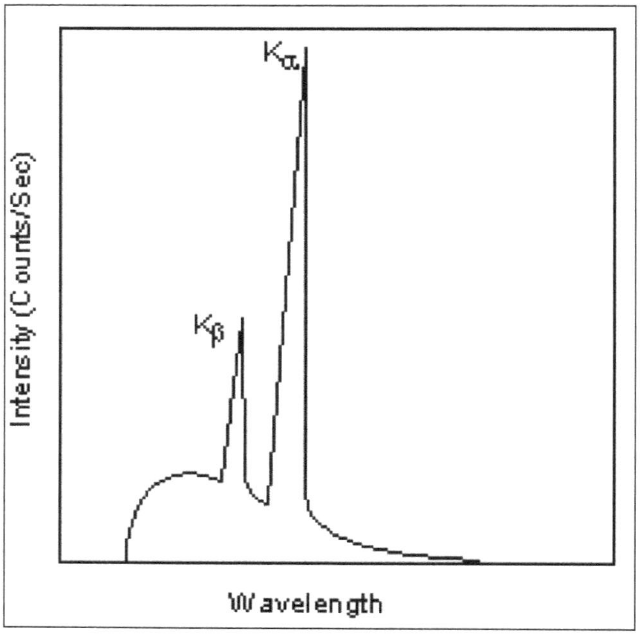

For commonly used target materials in X-ray tubes, the X-rays have the following well-known experimentally determined wavelengths:

Elements	K_α Wavelength (λ) Å
Mo	0.7107
Cu	1.5418
Co	1.7902
Fe	1.9373
Cr	2.2909

Principles of X-ray Diffraction/Crystallography

X-ray Reflection according to W. L. Bragg

Consider a set of N+1 equidistant atomic planes of spacing d, and a monochromatic plane X-wave falling on it at a glancing angle θ. It is assumed that each atomic plane reflects a very small fraction of the incident amplitude, small enough so that the weakening effect of this reflection on the incident amplitude may be neglected throughout the crystal. Under most angles of incidence, θ, the waves reflected from neighbouring planes will show a phase difference, and where all the reflected waves come together at great distance from the crystal, the superposition of these waves of systematically increasing phases will lead to a cancellation of amplitudes and to optical field zero.

There exists, then, only the transmitted wave. If, however, the phases of all the reflected waves arrive within less than one half wave-length phase difference, then all reflected amplitudes will build up together to an optical field in the direction of reflection, without any actual cancellations of contributions. Should all waves arrive in the same phase, then full re-inforcement of the waves takes place to an amplitude of N+1 times that of the single reflected wave.

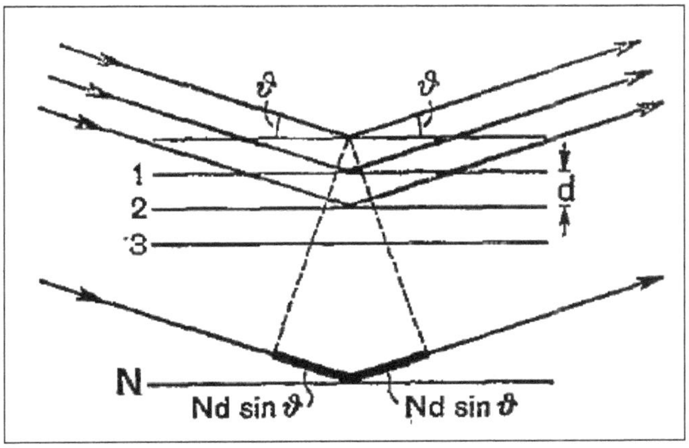

Bragg reflection on a set of N atomic planes.

Now the difference of optical path for the top and bottom wave is shown by the heavy-drawn path lying between two parts of the wave-fronts of the incident and reflected waves. Its length is 2Nd sin θ. The path difference between reflections on neighbouring planes is 2d sin θ. If this equals a whole number, n, of wave-lengths λ, then the phase difference is zero throughout the crystal. Therefore maximum amplitude of the reflected wave is obtained for angles θ_n such that:

$$2d \sin \theta_n = n\lambda$$

This 'Bragg Equation' determines the angles θn under which the first, second, third, . . ., etc. order reflections occur, for n = 1, 2, 3, . . ., etc. The greater the wave-length, the larger the glancing angle for reflection on the same plane; the greater the spacing, the smaller is the glancing angle for a given wave-length. If λ is known and θ_n measured, then the value of n/d follows, and if the order can be found, the value of the spacing of the set of reflecting planes is determined. By putting together the information on various sets of reflecting planes, obtained in this way with the X-ray spectrometer, the first crystal structure determinations were made.

As the angle of incidence is slightly varied from the Bragg angle, a phase difference develops between reflections from neighbouring planes and re-inforcement of the reflected waves becomes less perfect. Their effect cancels out to optical field zero if the maximum phase difference throughout the crystal corresponds to an entire wave-length path difference, or, indeed, to any multiple of it, say sλ, where s is an integer. For then there will be for any reflected elementary wave one of opposite phase superimposed. The condition for the angles under which these 'secondary zeros' of the reflection curve occur is:

$$2Nd \sin \theta_{n,s} = (Nn + s)\lambda$$

and the angular distance of the sth zero from the Bragg angle θn is given by:

$$2d \sin \theta_{n,s} - 2d \sin \theta_n = (s/N)\lambda$$

or, since the difference of angles is usually very small, this can be approximated by:

$$\theta_{n,s} - \theta_n = \Delta\theta_{n,s} = \tan\theta_n \cdot (s/N)$$

If we plot the reflected amplitudes as a function of the difference of the angles of incidence and reflection from the Bragg angle, $\varphi = \theta - \theta_n$, each Bragg angle is seen surrounded by zeros of the reflected amplitude and corresponding zeros of intensity obtained by squaring the curve. The principal or zero order maximum lies between the zeros for $s = \pm 1$ and has therefore a width of $(2 \tan \theta_n)/N$.

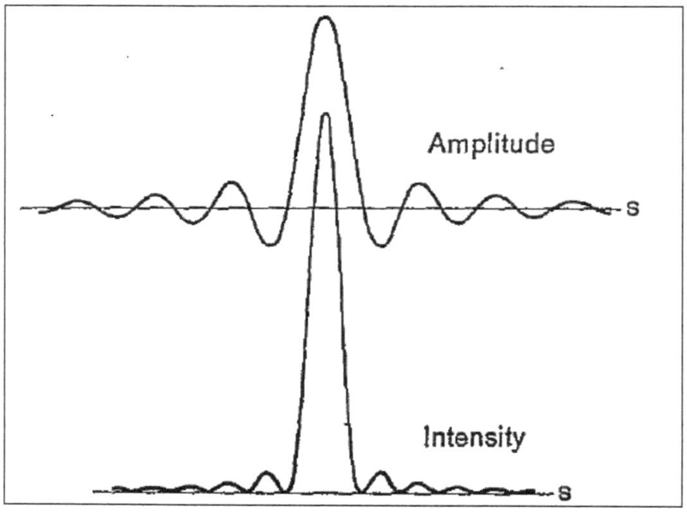

Amplitude and intensity curve of the wave reflected by a set of N atomic planes in the neighbourhood of the Bragg angle.

The finite angular width of the Bragg reflection is a consequence of the limitation of thickness of the crystal to Nd. The thinner the crystal, the wider is the angular range of the principal maximum, and the poorer is the resolving power of the crystal as an optical instrument for distinguishing between neighbouring wave-lengths in spectroscopy. The total reflected intensity in any order of reflection is proportional to the area under the principal maximum of the intensity curve. Since the amplitude maximum is proportional to N+1, or essentially to N if this is large, the maximum intensity is proportional to N^2. With the width of the principal maximum given above, the total reflected intensity is then proportional to $2N \tan \theta_n$. This shows that the reflected intensity is proportional to the thickness or volume of the crystal, - a result that is true only as long as the crystal is so thin that the reflection leaves the incident, and transmitted, ray at practically its full strength.

The consideration of the secondary maxima and of the finite resolving power of the crystal is essentially what H. A. Lorentz introduced into the theory, although using Laue's diffraction, not Bragg's reflection language. The measurement of the angular width of the principal maximum of reflection forms the basis of the determination of particle size by means of X-rays - in the particular case considered the thickness of a crystal flake could be determined from such measurements.

Bragg's formula can be applied to reflections on atomic planes which are not parallel to the crystal surface, because the condition of re-inforcement does not contain the orientation of these planes to the surface. If a small general refraction of X-rays in a crystalline medium is taken into account, λ and θ_n, in the formula have to be interpreted as the internal values which are related to the values outside the crystal by the ordinary laws of refraction. Since the refractive index for X-rays differs from 1 only by about one part in hundred thousand, this refraction need be considered only in measurements of very high precision; it was first found by W. Stenström of M. Siegbahn's school in 1919, and leads to a correction of wave-lengths or lattice constants when the usual Bragg formula is applied within the above range of accuracy.

The great simplicity of Bragg's theory is achieved by the introduction and use of the spacing d of the reflecting planes. Starting from the axial system or the cell of the crystal, the determination of d for different kinds of planes requires some geometry. On the surface Laue's theory appears more complicated, but it contains this internal geometry of the crystal as it was built-in. There were, besides, some objections to Bragg's idea of reflection which, in the early stages, made its acceptance not obvious. While densely populated planes of atoms are well defined, with wide spacings between them, the sparsely populated planes have little physical reality. Their spacing must be very small, in order to produce the fixed number of atoms per unit volume. Therefore it seems rather artificial to consider with Bragg that the few atoms on each plane reflect like a mirror, independently of the much closer atoms which belong to neighbouring parallel planes. Furthermore the regular arrangement of the atoms in any one of these planes is known, from the diffraction point of view, to produce many diffracted rays (cross-grating spectra), of which the specular reflection is only one.

X-ray Diffraction according to Laue

The Linear Grating

We first consider a row of equally spaced atoms, each of which becomes the source of a scattered spherical wavelet under the stimulation by an 'incident' monochromatic plane wave. The mode of propagation of a plane wave is represented by its 'wave-vector' k; this has length $1/\lambda$ (λ the given wave-length), and its direction is that of the normal to the planes of equal phase, or, as we may call these, the wave front. Let the three axial vectors by which the crystal is described be a_i (i = 1, 2, 3), then any point in space can be described by the 'coordinate vector' $x = x_1a_1 + x_2a_2 + x_3a_3$, where the xi are called the 'coordinate numbers'. It will be convenient to use the notion of the 'scalar product' of two vectors, for instance k·x. This is defined as meaning the product of the lengths of the two vectors (indicated by |k| and |x| respectively) times the cosine of the angle between their positive directions:

$$k\cdot x = |k|\cdot|x|\cdot \cos(k, x).$$

Now |x| cos(k, x) is the length of the projection of x on the direction of k; this projection is often called the resolved part of x along k. Evidently the resolved part is the same for all points x lying on the dotted line of figure, and, in three dimensions, for all points of the plane normal to k which contains the dotted line. Therefore the expression k·x = constant can be used to describe the planes of equal phase of a wave. The value of the constant is the length of the optical path, expressed in wave-lengths, from the wave front passing through the origin to that passing through point x. The argument $(-\nu t + k\cdot x)$ is characteristic of a wave of frequency ν travelling in the direction of k; $\nu/|k| = q$ is the velocity with which the phase travels.

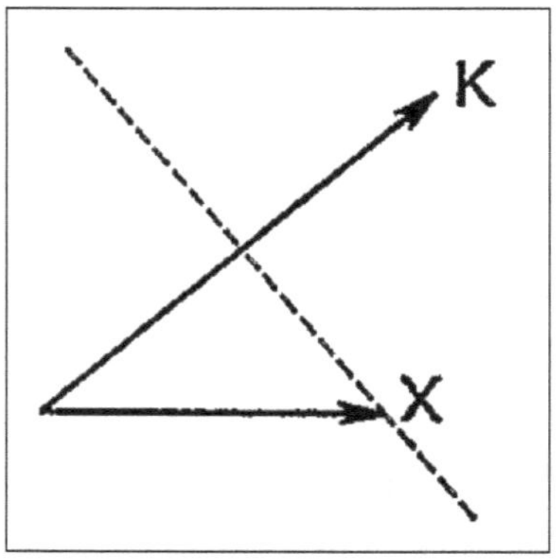

Scalar product of wave vector and coordinate vector.

Consider now the wavelets scattered by the .equally spaced atoms under the stimulus of the incident wave of vector k_1. Can a direction be found in which they all arrive in the same phase at a very distant point? We enumerate the atoms by an index l and call the vector from an atom l to its neighbour $l + 1$ the ‹translation' a. For full cooperation of *all* wavelets it is sufficient to find the condition for wavelets l and $l + 1$ to arrive at the point of observation without any difference of phase. In figure the wave fronts through atoms l and $l + 1$ respectively, are shown dotted. The optical path, measured in wave-lengths, through $l + 1$ is shorter than that through l by the resolved part of a along k, and longer by the resolved part of a along k_1. Therefore it is shorter by

$$k \cdot a - k_1 \cdot a = (k - k_1) \cdot a$$

For best re-inforcement this must be an integral number of wavelengths, say h wave-lengths. Thus a diffracted maximum occurs if k is such that:

$$(k - k_1) \cdot a = h$$

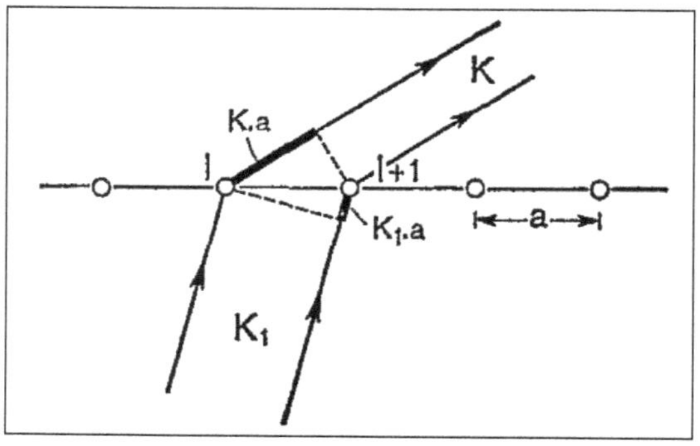

Diffraction by linear grating (row of atoms); difference of optical paths through neighbouring atoms.

By putting h = 0, a solution of this is seen to be $k = k_1$; that is, there always occurs optimum re-inforcement of the wavelets in the direction of incidence. To the right of this direction lie the directions with positive integers or '*orders*' h, to the left those for negative h. The values of h are limited by the condition that the diffracted waves have to move away from the row of atoms. The fact that the individual scattered wavelets are spherical will make the result of their superposition the same in all planes containing the row of atoms. The direction of k is therefore only the representative of a cone of directions surrounding the row, and maximum amplitude is achieved in all directions of observation along this cone. If we imagine the rays issuing from the row of atoms to be made visible by their intersection with a ground glass plate which is parallel to the row, the result will be a series of bright hyperbolae as in figure; the same pattern would be obtained on a glass plate placed beneath the row - or, in fact, on *any* plate held parallel to the row.

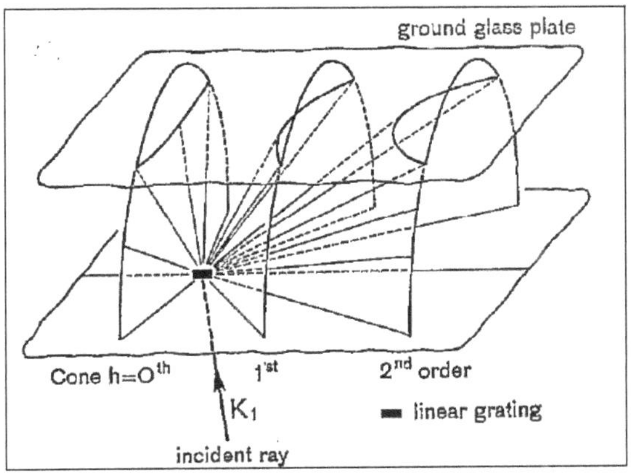

Diffraction by linear grating shown on ground glass plate in physical space.

To this first description of the directions of diflraction in relation to the row of atoms we now add a second description in a '*reciprocal space*'. We take from equation $((k - k_1)\cdot a = h)$ the following instruction for finding the direction of the wave vector k of a diffracted wave, given the direction of the wave vector k_1 of the incident wave: make the resolved parts of the two vectors with respect to the translation a such that they differ by a multiple (h) of $1/|a|$. To this end we construct a series of equidistant plahes normal to the row of atoms and with spacing $1/|a|$ between them. If the given wave vector k_1 is laid down so that it ends in the origin o, its starting point T is determined. If now from T we draw *any* vector k ending on one of the planes, say that labelled h, then condition is fulfilled and *all* wavelets arrive at a distant point in the direction of k in the same phase. But in classical scattering there is no change of wave-length; therefore $|k| = |k_1| = 1/\lambda$. This means geometrically that the end-point of k must also lie on a sphere about T of radius $1/\lambda$, i.e. which passes through the origin. This '*sphere of reflection*' is intersected by the set of planes in circles, and these are the geometrical loci for the ends of the diffracted wave-vectors. Connecting T, which will be called the '*tie-point*', to all points of the circle in the plane numbered h, one obtains the cone of diffracted directions of order h, as in figure.

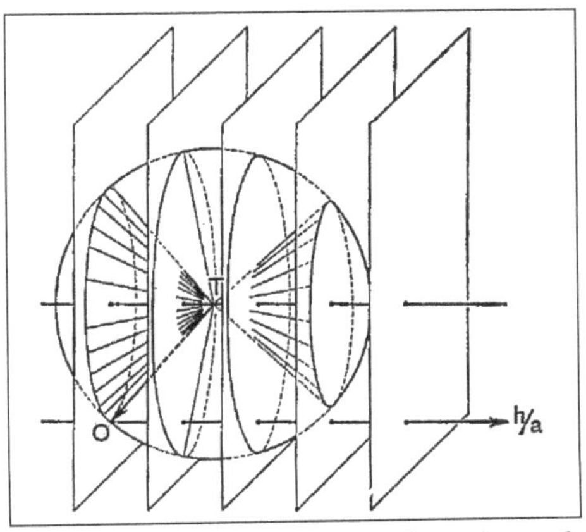

Diffraction by linear grating shown in reciprocal space.

In this construction the distances at which the planes are drawn represent $1/|\mathbf{a}|$, and the radius of the sphere of reflection is $1/\lambda$; thus we are using a space in which the distance represents the inverse of a length in physical space and is measured in cm^{-1} or 1/Å (= $10^8 cm^{-1}$). There is nothing more unusual in this than if we represent, for instance, a velocity [cm/sec], force [g·cm/sec²] or an electric field strength [volt/cm] by the length of an arrow in a suitably labelled space. The space we are operating in above is called 'reciprocal space' because the product of a length in this space and a length taken in physical space is a dimensionless quantity, i.e. a pure number.

We shall now extend the same two representations of the directions of diffraction, in physical space and in its reciprocal, to two- and three-dimensional lattices of atoms.

The Cross-grating

This is another name for an array of scattering centres filling a plane in a periodic manner. All atoms - if we take these to be the centres - are obtained from an original one by applying two translations, given vectorially by a_1 and a_2. If l_1 and l_2 are independent integral numbers, ranging from $-\infty$ to $+\infty$, the positions of the atoms are:

$$x_{l1,l2} = l_1 a_1 + l_2 a_2$$

The condition for maximum re-inforcement of *all* wavelets issuing from these points is that there be no phase difference between the wavelets coming from an atom and its two neighbours by the a_1 and a_2 translations, respectively. But this is exactly the condition applied twice over, once using the translation a_1 and an integer h_1, and again with a_2 and an independent integer h_2. Thus the condition for the wave-vector k of a diffracted wave is:

$$(k - k_1)\cdot a_1 = h_1$$

$$(k - k_1)\cdot a_2 = h_2$$

We visualize this condition with the help of the ground glass plate of figure which we imagine to be parallel to the plane of the cross-grating. Each of the above equations is fulfilled along the hyperbolic intersections of the plate with the cones surrounding the directions of a_1 and a_2. The directions for which both conditions are fulfilled are those leading to the points of intersection of the hyperbolae. Each diffracted ray is named as a two-digit 'order', (h_1, h_2), the integers of which tell the whole number of wave-lengths path difference of the wavelets scattered by an atom and its neighbours along a_1 and a_2. Cross-grating spectra always exist, for any wave-length or angle of incidence. Should a_2 be considerably larger than a_1, then the second set of hyperbolae have a much narrower spacing than the first, and the points of intersection mark out the first set very clearly. Starting out from the central spot which gives the direction of the incident ray and has order (0, 0), one can easily '*index*' the diagram by assigning each point its (h_1, h_2) values.

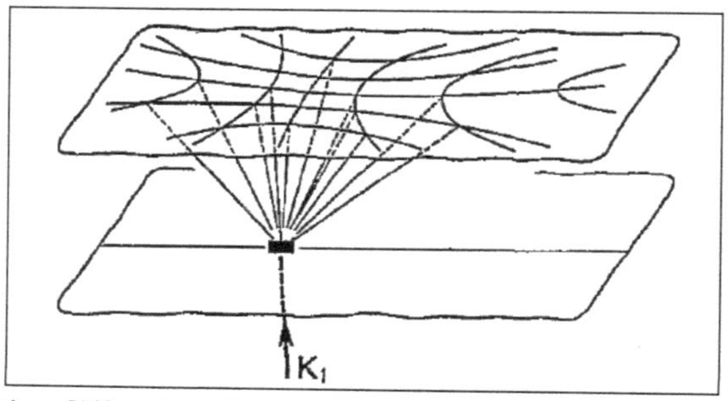

Cross-grating diffraction shown on ground glass plate in physical space.

The spherical wavelets combine equally on both sides of the cross-grating; a second ground glass plate beneath the grating would show the same design. To the transmitted beam on the upper plate corresponds a reflected beam on the lower plate, and this has again the order (0, 0). These two beams (0, 0) are *always* present, whatever the atomic distances, or even their regularity may be.

Let us now look at the corresponding construction in reciprocal space. The first of the two conditions leads to the set of planes normal to a_1 with spacings $1/|a_1|$ between them which we know from figure. The similar set of planes representing the second condition is normal to a_2, with spacing $1/|a_2|$. Both conditions are simultaneously fulfilled on the intersections of these sets, that is, on a periodic array of straight lines normal to the plane of a_1 and a_2. Additional to this, the condition of scattering without change of wave-length requires that the vector k of a diffracted wave end on the sphere of reflection of radius $|k_1|$ and centre at T. The wave-vectors $k_{h1,h2}$ of the diffracted waves therefore start out from T and end in the points of intersection of the sphere of reflection with the straight lines of orders h_1, h_2. Provided the radius of the sphere of reflection is not too small, intersections will always exist on the upper and lower hemisphere, corresponding to the cross-grating spectra of the same orders emitted to both sides of the cross-grating.

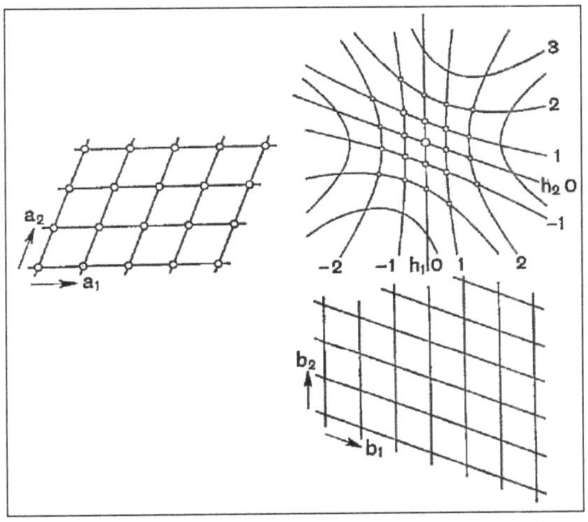

Left: cross-grating in physical space; upper right: pattern of cross-grating spectra in physical space; lower right: trace of the planes normal to a_1 and a_2 for construction in reciprocal space.

Figure illustrates the cross-grating with translations a_1 and a_2, and its relation to the pattern of crossed hyperbolae in physical space; it further shows in reciprocal space the ground plan of the sets of planes normal to a_1 and a_2, and the distribution of their lines of intersection. The area of the parallelograms between these lines can be transformed into a rectangle whose one side is $1/|a_1|$, while the other (vertical) side is $1/(|a_2| \sin \alpha)$, where α is the angle formed by a_1 and a_2. Thus the area is $1/(|a_1||a_2| \sin \alpha)$, and this is the inverse of the cell (or parallelogram) of the cross-grating.

Lattice in Three Dimensions

If cross-gratings, formed by two translations a_1 and a_2, are stacked on top of one another by a third translation a_3, not lying in the plane of a_1 and a_2, then a lattice is obtained with atoms lying at:

$$x_1 = l_1 a_1 + l_2 a_2 + l_3 a_3 \quad (l_i \text{ integers})$$

The condition for the diffracted rays from all cross-gratings to be superimposed with the same phase in the direction of a wave vector k is once again of the form. Therefore we have now to fulfill three conditions, containing three arbitrary integers h_i, namely:

$$(k - k_1)\cdot\mathbf{a}_1 = h_1$$

$$(k - k_1)\cdot\mathbf{a}_2 = h_2$$

$$(k - k_1)\cdot\mathbf{a}_3 = h_3$$

These are the conditions which Laue gave, in slightly less condensed form, in his first paper on the subject.

What does the third condition add to the first ones in the two geometrical representations? The ground glass plate of figure, which is parallel to the plane of a_1, a_2 will be intersected by a_3. For clarity we may assume the direction of a_3 to be sufficiently steeply inclined to a_1, a_2 so that the cones surrounding a_3, which represent the third condition, intersect the glass plate in ellipses; if a_3 is normal to the plate, the intersections will of course be circles. For all three conditions to be fulfilled simultaneously, an ellipse or circle must pass through an intersection of the hyperbolae. This will in general not occur, except for the direction of incidence, and then the incident wave will travel through the crystal without being diffracted. However, as we decrease the wave-length, the cross-grating pattern contracts towards its centre, the direction of the incident ray, and the ellipses or circles contract about the point where the axis a_3 intersects the plate. These contractions take place at different rates, and this means that for certain wave-lengths three curves intersect simultaneously and a diffracted ray of order (h_1, h_2, h_3) flashes up. This is the pass-filter action of the crystal for white X-radiation, since, given the direction of incidence, only a certain wave-length can appear in a diffracted ray, other wave-lengths not being admitted at this particular angle of diffraction. This statement has to be modified, however, because by multiplying the lengths of k and k_1, as well as the order numbers h_i by the same integral factor n, the equations (6) remain fulfilled without change of direction of k or k_1. In other words, if a wave-length λ is diverted by diffraction into a direction k, then λ/Z, $\lambda/3$, ..., etc. can be contained in the same direction by diffraction at correspondingly higher orders, provided, of course, these shorter wave-lengths are contained in the spectrum of the incident white radiation. This multiplicity of the wave-lengths in the spots of Laue diagrams makes these less suitable for crystal structure determination than the diagrams obtained with monochromatic radiation.

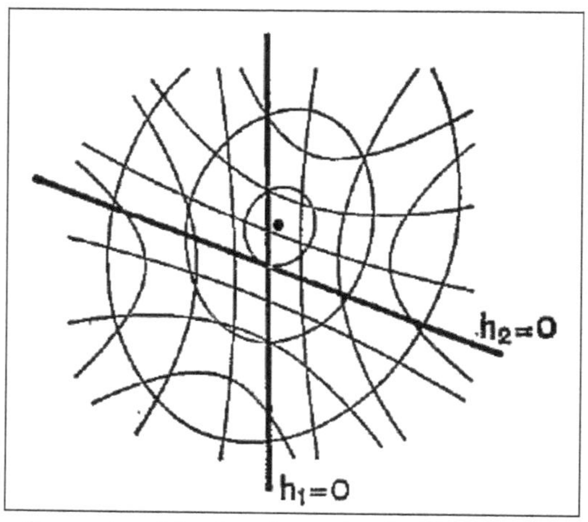

Diffraction by three-dimensional lattice shown in physical space on ground glass plate.
The ellipses represent the third condition of re-inforcement of the wavelets.

In reciprocal space, the third of the conditions adds a set of equidistant planes normal to a_3, and of spacing $1/|a_3|$, to the array of straight lines representing the first two conditions. The intersection of these lines or rods with the planes yields a three-dimensional lattice of points called the '*reciprocal lattice*'. The condition of diffraction is now that the sphere of reflection intersects a point of the reciprocal lattice. Again, with the direction of incidence and the wave-length given, - in short, given the tie-point -, the chances are that the sphere of reflection will not pass through *any* point of the reciprocal lattice, except the one at the origin through which it passes by definition. If this is so, only the 'primary' beam (000), the continuation of the incident beam, is formed and there is no diffraction in the usual sense. If, however, we vary the position of the tie-point in any way, the sphere of reflection will sweep over some of the lattice points, and every time such intersection occurs, a diffracted ray will be flashing up. In the Laue method the direction of incidence is fixed, but the length of the vector k_1 is variable, inversely as the wave-length; therefore the tie-point moves on a straight line passing through the point (000), and the radius of the sphere varies accordingly. In the methods working with monochromatic X-rays, T is free to move on a sphere of radius $1/\lambda$, the ‹wave number sphere'. If there is incidence in all directions with respect to the crystal lattice, as in powder diagrams, T moves over the entire surface of this sphere.

In wide-angle diagrams, T moves only within that part of the sphere that is cut out by the solid angle of incidence. In a rotation or oscillation diagram, finally, T is restricted to a great circle, or part of it, which is normal to the axis of rotation or oscillation. In each case, the process of ‹indexing' a diffraction diagram is essentially the reconstruction of this geometry from the diagrams, and it thereby leads to the determination of the shape of the reciprocal lattice. This process is made easier by any additional information obtained about the spots on the diagram. It is hardest in the case of powder diagrams, where nothing is known; the higher the crystal symmetry is known to be, the easier it becomes. Oscillation diagrams are easier to index than rotation diagrams on which more overlapping of spots may occur. It is of further great help to know the position of the crystal at the instant of reflecting each spot of a rotation or oscillation diagram. This is achieved in the so-called goniometer methods by coupling a displacement of the film to the rotation of the crystal and limiting the recording of spots to certain types of reflections by ‹layer-line diaphragms', as in the Weissenberg and Schiebold-Sauter goniometers. In the De Jong-Bouman and the Buerger precession camera film and crystal are moved in such a way that the spots appear in a pattern which is a section through the reciprocal lattice itself, so that there is no problem of indexing. In all cases, however, indexing is a routine procedure, once it has been done a few times.

The third part of figure which indicates the cross-section of the array of parallel rods normal to the plane of a_1, a_2. The same is shown in perspective in figure together with the third axis, a_3. Two successive planes normal to a_3 are shown (which are at distance $1/|a_3|$ from one another), and the points marked in which they intersect the rods. The eight intersections suspend a parallelopipedon which is the repeat unit or cell of the reciprocal lattice. Let us determine its volume. We know already that the ground plan of the cell had an area 1/A, where A was the area defined by the axes a_1 and a_2. The volume of the cell is obtained by multiplying this area with the height of the cell, OQ. But we know that the resolved part of OQ in the direction of a_3, that is, OP has length $1/|a_3|$. Therefore $OQ = (1/|a_3|) : \cos\gamma$, where γ is the angle POQ. Thus OQ is the inverse of the thickness of the crystal cell, namely $1/(|a_3| \cos\gamma)$, and this makes the volume of the cell of the reciprocal lattice the inverse to the volume of the crystal cell.

The faces of the reciprocal cell are formed of planes normal to a_1, a_2, and a_3, respectively. The edges of the cell are the translations of the reciprocal lattice, and we call them b_1, b_2, b_3; each of them is therefore normal to two of the a-axes. This fact can be expressed by stating that the scalar products vanish,

$$b_i \cdot a_k = 0 \text{ for } i \neq k$$

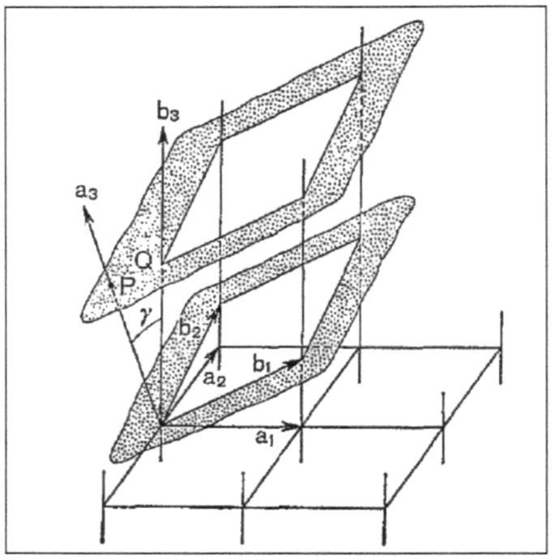

The reciprocal lattice in three dimensions. The origin O is the corner from which the vectors a_i and b_i are drawn. a_1 and a_2 are actually shown representing length $1/|a_1|$ and $1/|a_2|$, and OP = $1/|a_3|$. Then b_1 and b_2 are as shown and b_3 = OQ. These are the reciprocal vectors to the a_i, fulfilling Eqs. and they support the reciprocal lattice.

This property of the b-axes is shared by the 'polar axes' which Bravais introduced in the 1840's, but Bravais added an unsuitable definition of the length of the axes. For the reciprocal axes, which were first introduced by Willard Gibbs in his lectures at Yale University, the normalization can easily be seen from figure for b_3 = OQ. The result is that the resolved part of OQ in the direction of OP is the inverse of the length of OP; in other words, the scalar product of a_3 and b_3 has value 1. Since the decomposition of the crystal lattice into cross-gratings might have been performed with the choice of any two a-axes, analogous equations must hold for all axes, namely:

$$b_i \cdot a_i = 1 \; (i = 1, 2, 3)$$

The conditions (bi · ak = 0 for i≠k) and (bi · ai = 1 (i = 1, 2, 3)) can be condensed to the elegant statement of the relation between the a-axes and their reciprocals, the b-axes

$$b_i \cdot a_k = \delta_{ik}$$

Where, δ_{ik}, called the Kronecker symbol, signifies 1 if i = k and 0 if i ≠ k. In this form the equivalence of the two sets of axes appears very forcibly. Just as we obtain the b-set of axes from the a-set, so we construct the a-set from a given b-set. In crystallography, many of the optical or X-ray measurements lead in the first instance to the b-axes.

A lattice vector in the reciprocal lattice is defined as the vector from the origin to any other lattice point. Thus it is:

$$h = h_1 b_1 + h_2 b_2 + h_3 b_3$$

in analogy to Eq. ($x_1 = l_1 a_1 + l_2 a_2 + l_3 a_3$) for the lattice vector in the crystal lattice. We now show that h has the direction of the normal of the net plane in crystal space which has the Miller indices h_i. From the definition of these indices we know that the intercepts such a plane makes on the axes are as $|a_1|/h_1 : |a_2|/h_2 : |a_3|/h_3$, and, multiplying with $h_1h_2h_3$, they are as $h_2h_3|a_1| : h_3h_1|a_2| : h_1h_2|a_3|$. Thus $h_2h_3a_1$, $h_3h_1a_2$, $h_1h_2a_3$ are three points p_1, p_2, p_3 through which the plane with Miller indices h_1, h_2, h_3 may be laid. The vector h is normal to this plane if its scalar product with the two vectors $p_2 - p_1$ and $p_3 - p_1$, which lie in the plane, vanishes. The values of these products are found by multiplying term by term:

$$h \cdot (p_2 - p_1) = (h_1b_1 + h_2b_2 + h_3b_3) \cdot (h_3h_1a_2 - h_2h_3a_1)$$

$$h \cdot (p_3 - p_1) = (h_1b_1 + h_2b_2 + h_3b_3) \cdot (h_1h_2a_3 - h_2h_3a_1).$$

Because of the relations $(b_i \cdot a_k = \delta_{ik})$ both products vanish, and this proves the statement.

If in $(h = h_1b_1 + h_2b_2 + h_3b_3)$ the component numbers h_i have no common factor, we denote them by h_i^* and the lattice vector by h^*. This vector ends in the first point of a row of equidistant points whose positions are obtained by letting a common factor n of the component numbers h_i assume all integral values, positive and negative. This is indicated by writing:

$$h = nh^*$$

The linear lattice of spacing $|h^*|$ is, in reciprocal space, the image of the set of planes in physical space which are normal to h, and it is easy to show that their spacing is inverse to the spacing along the row of points along h:

$$d_h = 1/|h^*|$$

This spacing d_h is the one entering in the Bragg formula $n\lambda = 2d \sin \theta$.

To the vector $h = nh^*$ there corresponds the spacing d_h/n, or, in the Bragg formula, the *n*th order reflection on the set of planes.

The relation between the reflection and diffraction terminologies becomes evident in figure which shows a plane of the reciprocal lattice, the tie-point T and two diffracted rays. For one of these, the wave-vector k_h is entered. The fact that the sphere of reflection passes through the lattice point h can be stated in the equation:

$$k_h - k_1 = h \text{ (Laue Equation)}$$

This one-vectorial equation expresses the same facts that are contained in the three scalar equations $((k - k_1)\cdot a_1 = h_1, (k - k_1)\cdot a_2 = h_2, (k - k_1)\cdot a_3 = h_3)$.

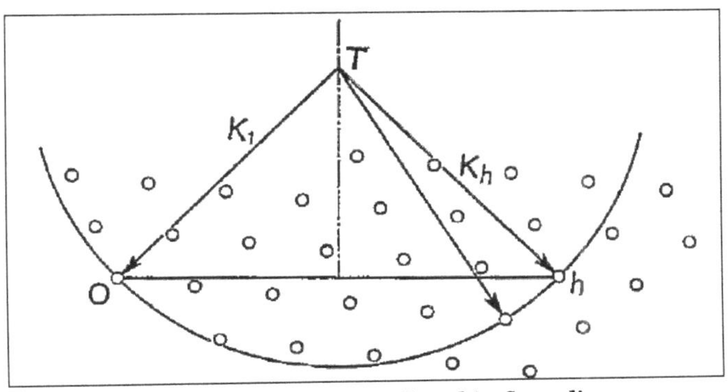

Relation between 'diffracted' and 'reflected' rays.

Because k_1 and k_h have the same length, the plane at right angles to the vector h and passing through its mid-point contains the tie-point. kh can thus be said to be the reflected image of k_1 by this plane - and this plane belongs to the set of atomic net planes with Miller indices (h_1, h_2, h_3). This shows that every one of Laue's 'diffracted' rays is simultaneously one of Bragg's 'reflected' rays.

Fourier Space

So far, the reciprocal lattice has been introduced as a convenient means for visualizing the directions under which alone the perfectly periodic crystal gives diffraction. Since in all these directions the scattered wavelets combine without phase difference, the amplitude generated by a finite crystal in these directions is proportional to N, the number of atoms of the crystal, and the energy, which is always proportional to the square of the amplitude, would be proportional to N^2, i.e. to the square of the volume of the crystal if the latter is 'bathed in X-rays. This is not the intensity which would normally be observed, because the larger the (perfect) crystal, the greater is its resolving power and the sharper its filter action, so that, with increasing N, the angular width of the principal maximum and the spectral range of admitted wave-lengths decreases. If we consider a finite, and therefore in the mathematical sense non-periodic crystal, there is a certain latitude in the fulfilling of the Laue-Bragg conditions, because for a very slight infringement the wavelets coming from the crystal will continue to re-inforce one another. This means that the vector k of the diffracted wave need not end on the lattice point of the reciprocal lattice, as long as it ends nearby, and that in this case the amplitude will be less than the maximum one. We can therefore plot the amplitude distribution in the space between the lattice points, similar to the amplitude distribution shown for the one-dimensional case in figure. The observed intensity may be taken as proportional to the integral value under the principal peak of the energy distribution, which is obtained by squaring the amplitude distribution. The limits of this peak are set by vectors k for which one extra wave-length path difference develops throughout the crystal, beyond that existing when k ends at the lattice point. For then the crystal can be divided into two halves, so that to each reflecting plane in the first half there exists one in the second half whose reflections are by $\lambda/2$ out of step with those of the first. If the areas of these corresponding planes were equal, full cancellation would occur; if they are unequal, only a surface effect remains over, instead of a volume effect.

Let us assume that the crystal is a parallelopipedon oriented according to the crystal axes, i.e. containing N_1 cells along a_1, N_2 along a_2, and N_3 along a_3. Considering the direction a_1, the path difference between neighbouring wavelets is $h_1\lambda$ for the diffraction of order (h_1, h_2, h_3). Therefore between the two extreme wavelets coming from the crystal the path difference is $N_1h_1\lambda$ if k ends at the lattice point (h_1, h_2, h_3). If we increase this path by sλ, where s is an integer, we get near-zero field. But this means that between any two neighbouring planes h_1 and h_1 + 1 of the reciprocal lattice, which have a distance $1/|a_1|$ between them, there are interleaved N_1 parallel planes which are the geometrical loci for those wave-vectors k for which the resultant amplitude is zero, or nearly so. The same can be done in the two other directions with corresponding spacings $1/(N_i|a_i|)$. We thus obtain a division of the cell of the reciprocallattice into $N_1N_2N_3$ subcells whose shape is reciprocal to the shape of the whole crystal. On the walls of these subcells end the vectors k for which the diffracted optical field nearly vanishes; in the interior of each subcell the diffracted amplitude will have a (positive or negative) maximum value, and the intensity distribution a positive maximum, but the height of these maxima falls off rapidly with increasing distance from the lattice point, because an ever smaller fraction of the crystal produces those wavelets which are not cancelled out by others in opposite phase. All this is but the extension of figure into three dimensions.

We have now filled reciprocal space with an amplitude distribution which is the same about each point of the reciprocal lattice. Whereas these points themselves are indicative of the crystal lattice, the subcell walls and the amplitude distribution contained between them have nothing to do with the inner periodicity of the crystal; they are, instead, determined by the external shape of the crystal. In fact, they would remain unchanged about the origin point (000) if the external shape were filled with an amorphous distribution of scattering matter - liquid or glassy -, while in this case the reciprocal lattice points outside the origin lose their significance, and with them the surrounding amplitude distribution. The construction of the zero amplitude subcell walls can be obtained easily by what is known in optics as Fresnel Zones.

Figure shows in a schematic way how the lattice points in reciprocal space are surrounded by the amplitude distribution (the 'shape transform' of the diffracting crystal). The division in the drawing is in twentieths and in tenths of the axes b_1 and b_2 and the shape of the crystal is indicated in the drawing. This corresponds to a crystal of 20 x 10 cells. The sphere of reflection intersects the distribution about the origin, and this would give rise to small angle diffraction. Besides, the sphere passes close by two other lattice points and the diffracted field in these orders will be given by the amplitude (or intensity) distribution on the surface of the sphere. In the upper intersection we see that the spot will be split into two parts, where the sphere intersects the two central rows passing through the point (h_1, h_2, h_3). These rows are principal peaks for either the h_1 or the h_2 directions, and as such have double width, as we know from the one-dimensional case of 6-1. Such ‹intensity spikes' always occur when the shape of the crystal contains relatively large flat faces, and their occurrence is easily understood by applying a Fresnel zone argument. Laue was the first to point this out and to explain the splitting of diffracted spots in electron diffraction pictures of very small and regular, electrolytically deposited metal crystals of octahedral shape. If the crystals are larger, the whole amplitude distribution contracts around the lattice points and becomes unobservable; the integrated value of the intensity remains measurable, and the integration produces the Lorentz factor.

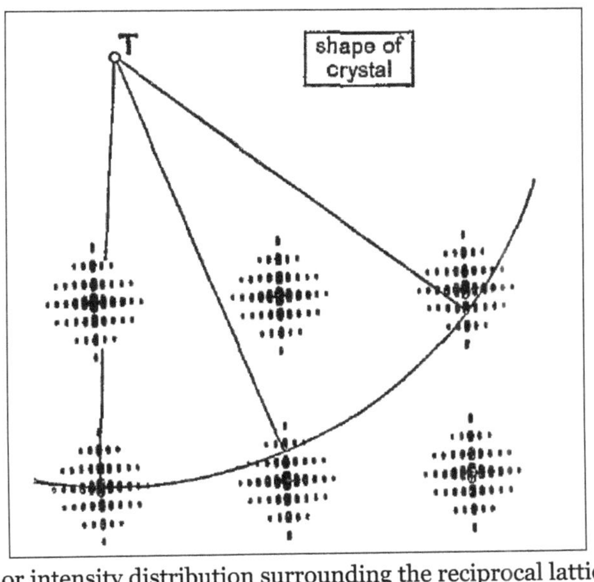

Amplitude or intensity distribution surrounding the reciprocal lattice points in the case of diffraction by a finite rectangular crystal (Crystal Shape Factor).

The previous may be taken as an example of the way in which continuous amplitude and intensity distributions can be plotted in the space surrounding the lattice points of the reciprocal lattice. Because this is very closely connected to the mathematical theory of Fourier Transformation, the space in which the reciprocal lattice is imbedded is best called Fourier Space. This gets rid of the term 'reciprocal space', which is a bad term because reciprocity is a symmetrical relation between two things and therefore unsuitable for designating one of them. The modern presentation of the (kinematical) diffraction theory is governed and simplified by the mathematical notion of Fourier transformation.

It can be shown that a distribution of matter, or electron density, or any other property throughout the crystal, can be described in two fundamentally different, but fully equivalent ways. One is, to give this function, say $\rho(x)$, point by point in physical space, that is, as a function of the position vector x (or its component numbers x_1, x_2, x_3) . The alternative one is to give, instead, a complete description of all the diffraction effects obtainable from this distribution of mass with all possible wave-lengths, and this means to give the value of the diffracted amplitude F throughout Fourier space, that is for all values of the position vector $\eta = \eta_1 b_1 + \eta_2 b_2 + \eta_3 b_3$ in this space. The transition from one description to the other is performed by integration and is therefore a straightforward mathematical operation to write down, though in many cases difficult to carry out. In the case that $\rho(x)$ is strictly periodic (which includes its extending everywhere to infinity) $F(\eta)$ is zero except at the points of the reciprocal lattice where $\eta = h$; this may be expressed by saying that Fourier space is reduced to an 'index space', namely the points of the reciprocal lattice corresponding to the periodicity of $\rho(x)$. In that case the integral over Fourier space degenerates to a sum, or the Fourier integral to a Fourier series.

Bragg's Law

Bragg's law means that the diffraction can occur only when the following equation is satisfied.

$$n\lambda = 2d \sin\theta$$

Where,

- n is a positive integer,
- λ is the wavelength of the X-ray,
- d is the distance between the lattice plane,
- θ is the incident glancing angle (supplement of the incident angle).

One may think: "isn't θ the "incident angle?", but the "incident angle" is defined as the angle between the normal direction of a reflection plane and the incident ray (or more generally quantum beam), in a traditional theory of optics.

The Bragg's law is often expressed in another way:

$$\lambda = 2d \sin\theta$$

and this expression may be more popular than that of Eq. ($n\lambda = 2d \sin\theta$). Be sure that the left side of Eq. ($\lambda = 2d \sin\theta$) does not include "n".

The definition of d in Eq. ($\lambda = 2d \sin\theta$) is different from the definition of d in Eq. ($n\lambda = 2d \sin\theta$), and it is equivalent with (d / n) for the definition of d in Eq. ($\lambda = 2d \sin\theta$). In the following sections in this chapter, the expression by Eq. ($n\lambda = 2d \sin\theta$) will be used, because it looks a little easier to be understood.

Illustrate the Bragg's law

It is not a bad idea just to memorize the Bragg's equation, because it is often used on analysis of diffraction data. But such memory will easily be volatilized, if you do not use the equation for a while. And it is likely that you might forget the meanings of the symbols, even if you do remember the equation. It is recommended to draw illustrations several times, and to remember how the Bragg's is derived.

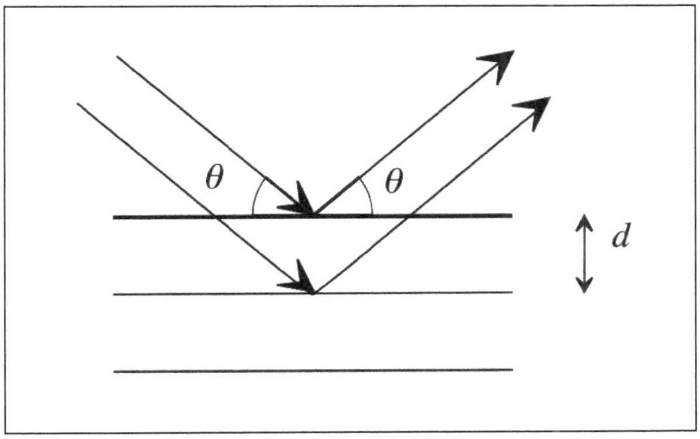

Illustration to derive the Bragg's law.

Difference in Path Length

The most significant point of the Bragg's law can be explained by the constructive interference, which occurs when the path difference of traveling waves matches with the integral multiplication of the wave length.

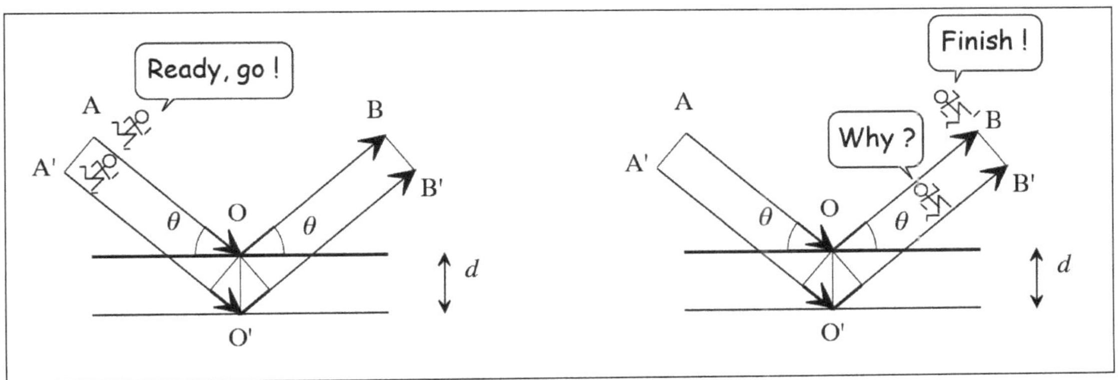

It will take additional length when you move along the path A'O'B' as compared with the path AOB. How long is the additional length (path difference)?

Look at the figure below and note that the path difference corresponds to the length of the two segments drawn as heavier lines.

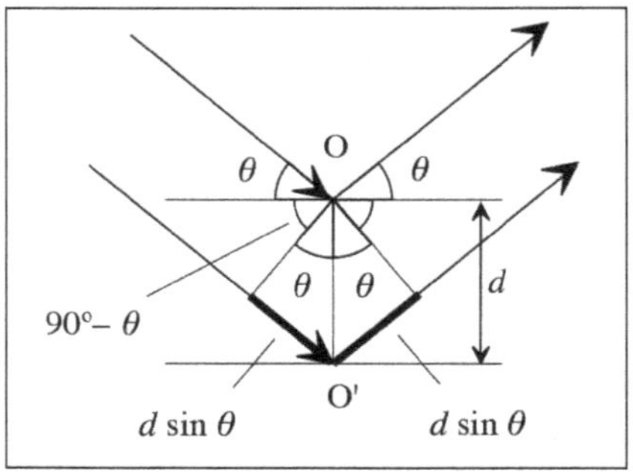

Relations among path length, d and θ.

The length of one segment should be " d sin θ ", because the segment is the opposite side to "the corner with the angle θ " in a "right triangle with the oblique side of length d ".

You may feel as if most part of the Eq. ($n\lambda = 2d \sin\theta$) have already been explained by the above relation, but actually, it can explain only small part of what the Bragg's law means.

Symmetric Reflection

The Bragg's law implies that the incident and reflected glancing angles are both equal to θ. Why should the angles of incident and reflected beams be coincided? Of course, the reflection by a mirror should be symmetric, but we are talking about the diffraction of X-ray by a lattice plane. The lattice plane should be something like planar arrangement of atoms.

What will occur on the asymmetric reflection shown in the figure below, for example?

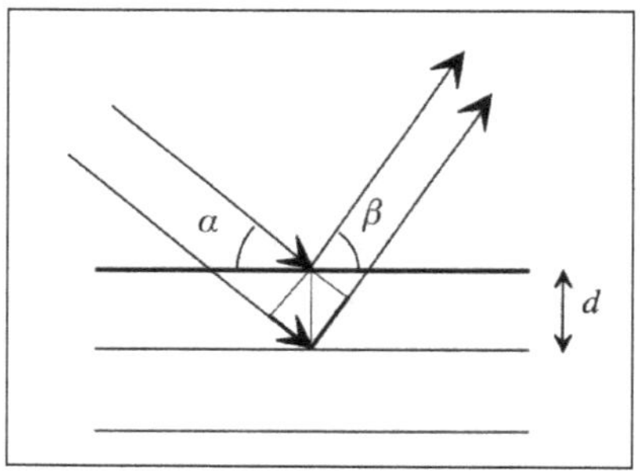

What will occur on the reflection with the incident (glancing) angle of α and reflected (glancing) angle of β?

The path difference should be " $d(\sin\alpha + \sin\beta)$ ", because it is also the sum of the two segments, the lengths of which are " $d \sin\alpha$ " and " $d \sin\beta$ ", as shown in the figure.

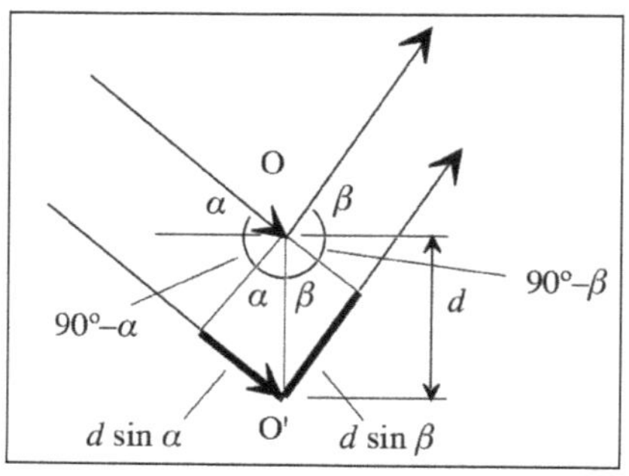

Path difference for the incident (glancing) angle of α and reflected (glancing) angle of β should be "$d(\sin\alpha + \sin\beta)$".

Don't you think diffraction may occur, if the value of "d (sin α + sin β) " matches to the integral multiplication of the wave length ?

Actually, diffraction does never occur in such a case. Why is the asymmetric reflection forbidden?? Now, the path difference in the asymmetric reflection appears to be "d (sin α + sin β)", because the reflection point O' on the second lattice plane is located right below the reflection point O of the first lattice plane. But the difference between the path reflected at O and O' should be "d [sin (α +φ) + sin (β −φ)]/cosφ", if the reflection point O' is located at the point displaced by such distance given by "d tanφ".

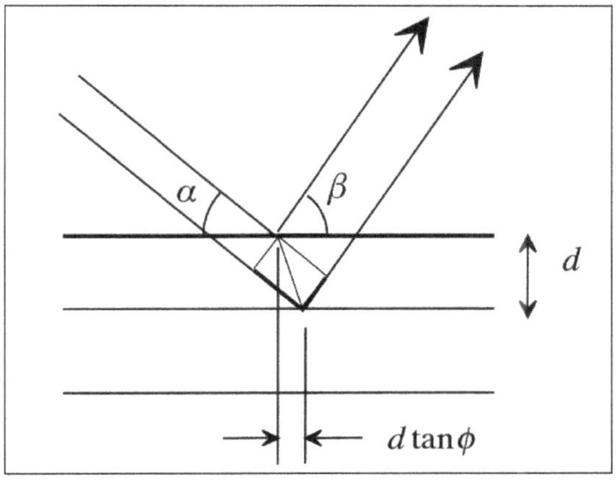

How about displacing the reflection point for asymmetric reflection?

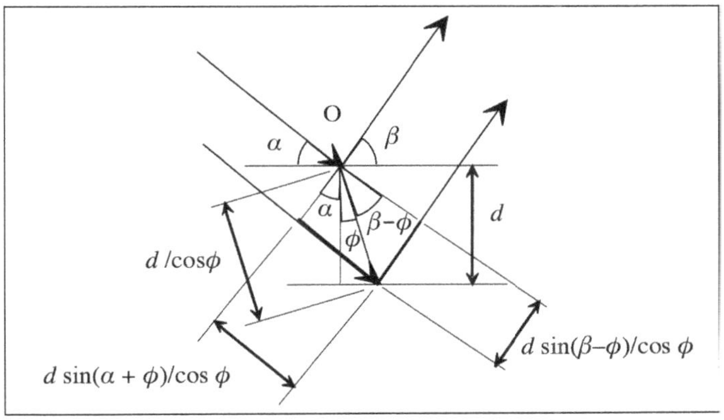

Path difference can be "d [sin (α +φ) + sin (β - φ)] / cosφ" for asymmetric reflection.

Both constructive and destructive interferences, depending on the location of the reflection point O' (different value of φ) can occur in the case of asymmetric reflection. There may be a special arrangement of atoms and angles α and β that lead only constructive interference, but it will be just a "symmetric Bragg reflection" for different lattice planes.

On the other hand, the path difference should always be "2 d sin θ" for symmetric reflection, no matter where the reflection point is located on a lattice plane. You can confirm this relation by the fact that the following equation holds for arbitrary φ,

$$\frac{\sin(\theta-\phi)+\sin(\theta+\phi)}{\cos\phi}=2\sin\theta$$

Which can easily be derived by applying trigonometric addition formulas. In conclusion, we can say that diffraction occurs only when the reflection is symmetric about the lattice plane.

Bragg's Condition

Furthermore, the Bragg's law does not only mean the constructive interference on Eq. ($n\lambda = 2d \sin\theta$), but it has a much more strict meaning to forbid any reflections not satisfying Eq. ($n\lambda = 2d \sin\theta$). How can this strict condition be derived?

So far, we have considered only two lattice planes, the top one and the 2nd one from the surface, but we must consider the 3rd, 4th, ...,etc. and much more lattice planes to explain the strict condition.

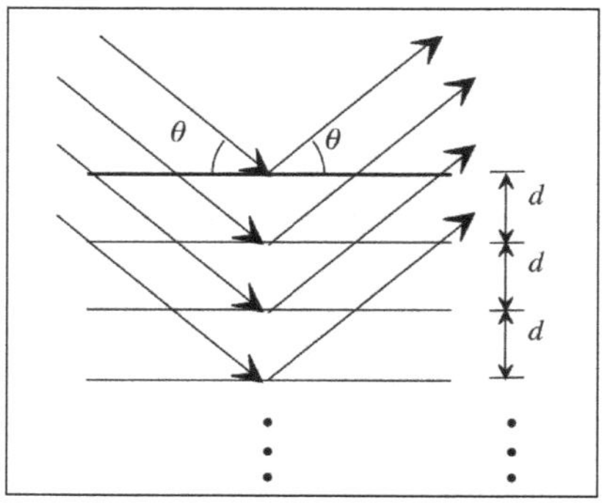

We must take many lattice planes into consideration to understand the Bragg's law.

The path difference is " $2d \sin \theta$ " for the reflection from the 2nd lattice plane, and it should be " $4d \sin \theta$ " for the 3rd plane, " $6d \sin \theta$ " for the 4th, and " $2(j-1)d \sin \theta$ " for the j-th lattice plane.

All the reflections from any lattice planes (1st, 2nd, 3rd, planes) will always be interfered constructively, if the relation:

$$n\lambda = 2d \sin \theta$$

is satisfied. Then, how about the case:

$$(n + 1/2)\lambda = 2d \sin \theta,$$

Which means the path difference is given by half-integer multiplication of the wavelength? The reflection from the 1st and 2nd planes will be canceled, and the reflection from the 3rd and 4rh planes will also be canceled, and so on. Most of the reflections will vanish except the last one for the odd number of planes.

Similarly, when the following relation:

$$(n + 1/3)\lambda = 2d \sin \theta$$

is satisfied, the sum of the reflections from the 1st, 2nd and 3rd planes will be canceled, the sum of the reflections from the 4th, 5th and 6th planes will be canceled, and so on. In general, when the following relation is satisfied,

$$(l/m)\lambda = 2d \sin \theta$$

Where (l/m) is an irreducible fraction.

The sum of the reflections from the 1st to the m-th planes are canceled, and the sum from (m + 1) to (2m) will be canceled, and so on. In conclusion, when the value of (2d sin θ) / λ are not an integer, there will always be a combination of planes which cancels the amplitude of the reflected waves. Therefore, diffraction will never occur, if Eq. ($n\lambda = 2d \sin\theta$) is not satisfied, in case of infinitely large number of lattice planes. (In other words, diffraction may be observed, even if the situation is slightly deviated from the condition defined by Eq. ($n\lambda = 2d \sin\theta$), in case of finite number of lattice planes.

Difference of Observation from Theory

Let us examine that we can certainly apply the Bragg's law to diffraction intensity data observed with a realistic instrument.

In a realistic diffraction measurement, the sample is irradiated by x-ray emitted from a generator, and the intensity of the diffracted beam is measured with a detector. Both the distance between the generator and the sample and the distance between the sample and the detector are several ten m, 20 cm = 200 mm, for example. We may apply the Bragg's law, if we can treat the distance "infinitely long".

Since the interference of X-rays, no matter constructive or destructive, can occur only when they are emitted from the identical point on the radiation source and also detected at the same point on the detector, the path difference should be expressed somehow shown in the figure below. Let us calculate the deviation of the realistic situation from that assumed in the theory.

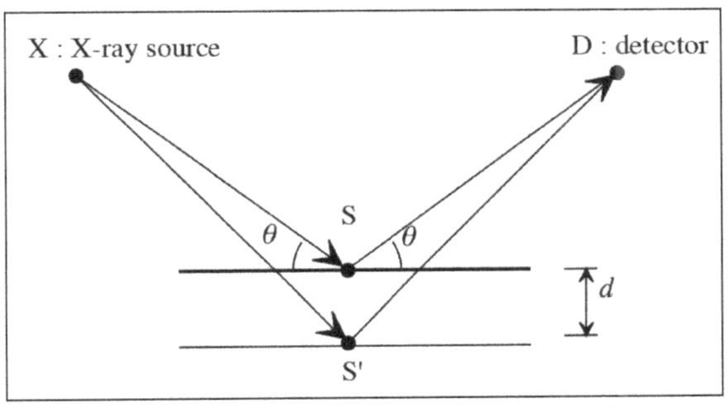

Exercise: evaluate the difference in length of the path XS'D and XSD, and compare it with the value of 2d sin θ, assuming XS = SD = R = 200 mm, d = 0.2 nm, θ = 21°.

As interatomic distance is usually 0.1 nm to 0.3 nm, assume that the distance of lattice plances is 0.2 nm, in this exercise. The angle to satisfy the Bragg's equation for n = 1, the interplanar distance d = 0.2 nm and the wavelength λ = 0.15 nm will be θ = 21°.

The length to be evaluated is given by 2 (R' - R) = 2 (XS' - XS) in the figure.

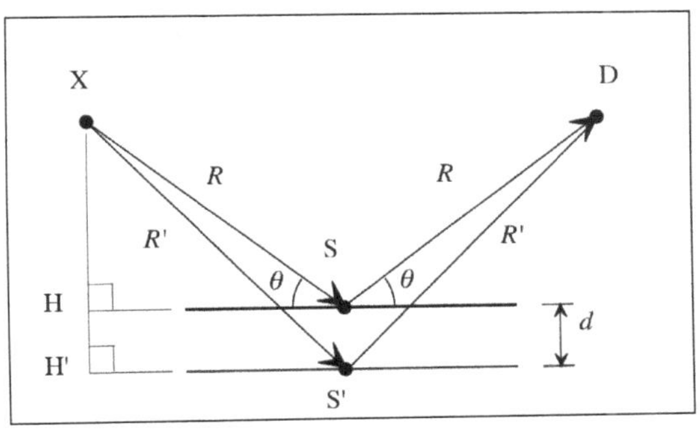

Let H and H' be the feet of the perpendicular lines from X to the two lattice planes.

Following relations are derived from figure,

$$XH = R\sin\theta,$$
$$XH' = XH + HH' = R\sin\theta + d,$$
$$SH = S'H' = R\cos\theta,$$

Then,

$$R' - R = \overline{XS'} - \overline{XS} = \sqrt{\overline{XH'}^2 + \overline{S'H'}^2} - \overline{XS}$$
$$= \sqrt{(R\sin\theta + d)^2 + (R\cos\theta)^2} - R$$
$$= \sqrt{R^2 + 2Rd\sin\theta + d^2} - R$$

It should be noted that the formula:

$$\sqrt{R^2 + \delta} - R \ (\delta << R^2)$$

is not suitable for calculation, even if you can use a calculator or a computer. It is well known that

it is better to use the formula on the right side of the following equation for numerical calculation,

$$\sqrt{R^2 + \delta} - R = \frac{\delta}{\sqrt{R^2 + \delta} + R}$$

Then, it should be calculated applying the following transformation,

$$R' - R = \sqrt{R^2 + 2Rd\sin\theta + d^2} - R$$
$$= \frac{2Rd\sin\theta + d^2}{\sqrt{R^2 + 2Rd\sin\theta + d^2} + R}$$
$$R = 0.2\ [\mathrm{m}]$$
$$2Rd\sin\theta + d^2 = (2R\sin\theta + d)d$$
$$= (2\times 0.2\times \sin 21° + 0.2\times 10^{-9})\times 0.2\times 10^{-9}$$
$$= 2.866943600362402\times 10^{-11}\ [\mathrm{m}^2]$$

for the exercise here, and we will have:

$$2(R' - R) = \frac{2\times 2.866943600362402\times 10^{-11}}{\sqrt{0.2^2 + 2.866943600362402\times 10^{-11}} + 0.2}$$
$$= 1.433471799924345\times 10^{-10}\ [\mathrm{m}]$$

On the other hand, we have:

$$2d\sin\theta = 2\times0.2\times10^{-9}\times\sin21^{\circ}$$
$$=1.433471798181201\times10^{-10}\ [\text{m}]$$

So, it is found that the two values are coincided up to 9 significant figures. If required, we can evaluate the difference of the two values by calculating,

$$2(R'-R)-2d\sin\theta=2\sqrt{R^2+2Rd\sin\theta+d^2}-2R-2d\sin\theta$$
$$=\frac{2\left[\left(R^2+2Rd\sin\theta+d^2\right)-\left(R+d\sin\theta\right)^2\right]}{\sqrt{R^2+2Rd\sin\theta+d^2}+R+d\sin\theta}$$
$$=\frac{2d^2\cos^2\theta}{\sqrt{R^2+2Rd\sin\theta+d^2}+R+d\sin\theta}$$
$$=1.743144824852707\times10^{-19}\ [\text{m}]$$

In conclusion, we can expect that the effect of the deviation of the observation from the Bragg's law to be extremely small and negligible. But it also shows that a kind of approximation is necessary for the Bragg's law to be applied to realistic data. It is recommended to calculate the difference by you, applying the values for the dimensions of the instrument you use for the experiment, and examine if it is certainly negligible on application of the Bragg's law.

Electron Crystallography

Electron crystallography is the quantitative use of different information by electron scattering to study perfect crystal structures as well as defects and interfaces.

Most properties of matter depend on its structure. Many different techniques have been developed over the centuries for structure analysis of matter. Crystallography is the most important technique for obtaining the complete atomic structure of matter in a crystalline phase. Among the three main radiation sources used in crystallography, X-ray is the dominant one. X-ray crystallography was founded in 1912, and is still today the most important technique for studying atomic structures of crystalline materials. Electron diffraction (ED) of single crystals was discovered fifteen years later and thereafter the wave property of electrons was exploited in the invention of the electron microscope. Since then, electron microscopes have been used in many fields as a tool for exploring and visualizing the microscopic world in all its beauty. Between the first electron microscopy (EM) images by Ruska 1933 and today's high resolution electron microscopy (HREM) images with atomic resolution lies seventy years of untiring engineering. More recently, the unprecedented power of computers has made it possible to analyze quantitatively, and even further improve, the HREM images. The amalgamation of quantitative electron diffraction and high (atomic) resolution electron microscopy with image processing has created a new powerful tool for structure analysis - electron crystallography. Modern analytical transmission electron microscopes now combine structure analysis by ED and HREM with chemical analysis by energy dispersive spectroscopy (EDS) and electron energy loss spectroscopy (EELS). This makes electron microscopy a powerful and complete analytical tool for studying materials.

Development of Electron Crystallography

Electron Diffraction Analysis Started Early in Moscow

Structure analysis of crystals by electron diffraction (ED) was first pursued in 1937-1938, by a group of crystallographers in the former Soviet Union, led by Pinsker and Vainshtein. For twenty years, they used their own designed electron diffraction cameras with relatively low accelerating voltages (<100 kV) to collect electron diffraction data of a large number of materials, such as basic salts, metal nitrides and carbides, semiconducting alloys and clay minerals, as well as some organic crystals. Most of the materials were oblique textured and gave electron diffraction of texture patterns. Using this electron diffraction data, they developed electron diffraction into a complete and independent structure analysis method. They addressed various problems which then could not be solved by X-ray diffraction, such as location of hydrogen atoms in crystals.

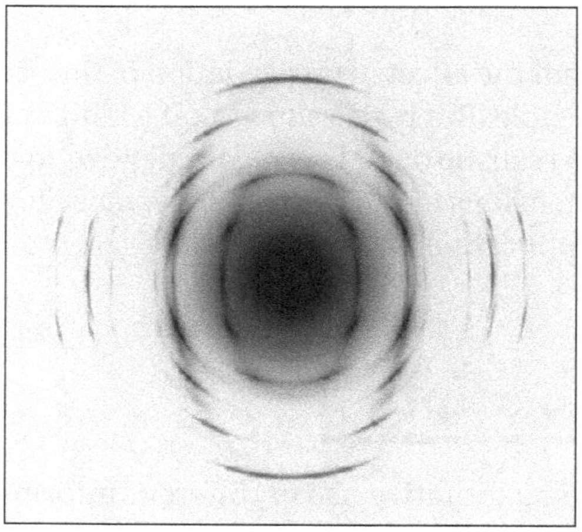

An oblique texture pattern of lizardite.

The method of structure analysis developed by the Soviet group was based on the kinematical approximation that ED intensity is directly related (proportional) to the square of structure factor amplitudes. The same method had also been applied by Cowley in Melbourne for solving a few structures. In 1957 Cowley and Moodie introduced the n-beam dynamical diffraction theory to the scattering of electrons by atoms and crystals. This theory provided the basis of multi-slice calculations which enabled the simulation of dynamical intensities of electron diffraction patterns, and later electron microscope images. The theory showed that if dynamical scattering is significant, intensities of electron diffraction are usually not related to structure factors in a simple way. Since that day, the fear of dynamical effects has hampered efforts to analyze structures by electron diffraction.

Experimental electron diffraction amplitudes, just as X-ray diffraction, do not contain the phase information of the structure factor which is needed for solving unknown structures. Various methods, such as the Patterson method and different trial and error techniques, were used for finding phase information in X-ray crystallography. These methods were used by the Soviet group also for phasing electron diffraction data. In 1953, Hauptmann and Karle introduced the so-called direct methods, for solving the phase problem. The direct methods, combined with the development of computers, accelerated very much the development of X-ray crystallography. In 1976, Dorset and Hauptmann in Buffalo for the first time applied the direct methods to electron diffraction data. This pushed structure analysis by electron diffraction a significant step forward. Using the direct phasing methods, Dorset has successfully phased electron diffraction data of various organic and inorganic crystals. These include aromatic molecules, lipids and other linear molecules and polymers, as well as some of the inorganic crystals previously collected by the Soviet group. His work has shown that electron diffraction, just as X-ray diffraction data, can be used for ab initio crystal structure determination.

Structure determination of unknown crystals by electron diffraction was performed by several research groups, on Al-Fe alloys by Gjønnes et al., on metal-cluster compounds by Weirich et al. and on zeolites by Wagner et al. Selected area electron diffraction or electron diffraction collected by a precession technique were used and the structure factor phases were deduced by direct methods, Patterson method or from convergent beam electron diffraction.

Electron Microscopy can be Combined with Image Processing

Quantitative electron microscopy (EM) for structure analysis can be traced back to the work by Aaron Klug and co-workers at the MRC laboratory of Molecular Biology in Cambridge in the 1960-ies. Klug started to study viruses by electron microscopy, soon after the double helix of DNA and the first protein structures were solved in that laboratory. His group developed optical diffraction and computerized image processing and first demonstrated 3D reconstruction by electron microscopy. This 3D reconstruction method is based on the fact that both phase and amplitude information are present in electron microscopy images and can be extracted from the Fourier transform of images by digitized image processing. They discussed various ways of determining the crystallographic structure factor phases which are needed for solving crystal structures and point out for the first time that these phases can be extracted directly from the Fourier transform of an EM image.

The 3D reconstruction method introduced by DeRosier and Klug created a revolution in structural molecular biology. Hundreds of macromolecular structures, including membrane proteins and viruses, were determined by this method. This method was further developed by Henderson and Unwin, and they determined first the 3D structure of a membrane protein, bacteriorhodopsin, to atomic resolution. Since then several other membrane structures, including PhoE porin and light harvesting complex, have also been solved to near atomic resolution (3-4 Å). Due to his development of crystallographic electron microscopy and his structural elucidation of biologically important nucleic acid-protein complexes, Aron Klug was awarded the Nobel Prize in Chemistry in 1982, the only Novel Prize on electron crystallography.

The resolution of EM images of crystals of biological macromolecules is limited by large unit cells, poor crystal ordering and radiation sensitivity. Most inorganic crystals give much higher resolution than biological samples do, due to smaller unit cells, better ordering and less radiation sensitivity. It is relatively easy to get high resolution electron microscopy images and diffraction patterns from them. The first EM images near atomic resolution (4Å) showing details within a unit cell were obtained from a thin crystal of $Ti_2Nb_{10}O_{29}$ by Ijima, then working at Arizona State

University. It was found that the contrast of HREM images changed with optical conditions and crystal thickness. In order to interpret the contrast changes, O'Keefe in Melbourne developed the multi-slice calculation method to simulate series of images of $Ti_2Nb_{10}O_{29}$ under different defocus and thickness condition and successfully interpreted the contrast changes in the experimental images. The success of image simulation led many microscopists to study the effects of the contrast transfer function and specimen thickness on images. Soon a consensus was reached that experimental HREM images can never be directly interpreted and used for solving atomic structures, but have to be confirmed by image simulation.

There are several disadvantages with the image simulation method. A nearly correct structure model is needed beforehand. This is often not available, especially for relatively complicated structures. Images are compared visually and no quantitative figure of merit is used for judging how well images and simulations agree.

The crystallographic electron microscopy developed by Klug et al. has also great impacts on inorganic crystals. Klug applied the crystallographic image processing method to a through-focus series of HREM images of a $GeNb_9O_{25}$ crystal and corrected for the contrast changes caused by the contrast transfer function. They showed that the correct structure projection could be retrieved from every single image in the series. This was the first attempt to interpret images of inorganic crystals without simulation.

The crystallographic image processing method was then used by Hovmöller's group in Stockholm for solving crystal structures of a number of niobium oxides. In the projected potential maps obtained by crystallographic image processing from experimental HREM images at 2.5Å resolution, all heavy atoms could be resolved. Atomic positions could be determined with an accuracy of about 0.1 Å. In the meantime, Li Fan-hua and Fan Hai-fu's group in Beijing developed the image processing method, combined with image deconvolution and phase extension by maximum entropy and direct methods, to solve the crystal structure of $K_2Nb_{14}O_{36}$. They also applied the direct methods for solving unconventional crystal structures such as quasicrystals and incommensurately modulated structures.

Since 1990, more and more structures have been solved from HREM images and electron diffraction and more and more scientists have become interested in structure analysis by electron crystallography. This pushed electron crystallography a big step forward. The term "electron crystallography" was first introduced and soon been accepted.

Combining HREM Images and Electron Diffraction

In HREM images of inorganic crystals, phase information of structure factors is preserved. However, because of the effects of the contrast transfer function (CTF), the quality of the amplitudes is not very high and the resolution is relatively low. Electron diffraction is not affected by the CTF and extends to much higher resolution (often better than 1Å), but on the other hand no phase information is available. Thus, the best way of determining structures by electron crystallography is to combine HREM images with electron diffraction data. This was applied by Unwin and Henderson to determine and then compensate for the CTF in the study of the purple membrane.

Electron diffraction has also been used for extending the resolution of images, using the direct methods from X-ray crystallography. Another method, based on the maximum entropylikelihood technique was introduced by Bricogne and Gilmore. The maximum entropy-likelihood method has been used for phase extension of electron diffraction data. Two successful applications are the extension of HREM image resolution of the organic structure perchlorocoronene from 3.2 Å to 1 Å and the 2D projection of the membrane protein bacteriorhodopsin.

Several other techniques, such as electron holography and convergent beam electron diffraction (CBED) have also been developed for structure analysis. CBED can provide information not only on the lattice parameters and the symmetry of crystals, but also accurate structure-factor amplitudes and phases. Accurate structure factor determination by CBED can provide information on the location of valence electrons. However, it is more favourable for thick crystals (> 500 Å) with small unit cells (< 10 Å).

Electron Crystallography have some Advantages over X-Ray Crystallography

X-ray crystallography is still the best technique for complete and accurate determination of crystal structures. However, electron crystallography has several important advantages over X-ray crystallography:

- Electrons interact with matter much stronger than X-rays. Thus much smaller crystals than those needed for single crystal X-ray diffraction can be analyzed by electron crystallography. For single crystal X-ray diffraction, crystals should have a size larger than $100\times100\times100$ μm^3 (or $10\times10\times10$ μm^3 on a synchrotron). For electron diffraction, crystals can be as small as $10\times10\times10$ nm^3. Even smaller crystals, down to some 10-20 unit cells, can be studied by HREM imaging. Electron crystallography has a promising future for structure analysis of crystals too small for X-ray diffraction analysis, such as grain boundaries, metastable phases etc.

- Electrons can be focused by magnetic lenses to form an image. High resolution electron microscopy images of crystals can be obtained, while Xray imaging is not possible. The phase information which is lost when registering diffraction patterns is preserved in HREM images.

- The mechanism by which electrons interact with crystals is different from that of X-rays. X-rays detect electron density distribution in crystals, while electrons detect electrostatic potential distribution in crystals. Electron crystallography may be used for studying some special problems related to potential distribution such as the oxidation states of atoms in the crystal.

- Almost all crystals suitable for X-ray powder diffraction can be studied by electron diffraction. Several of the most demanding problems with powder diffraction are overcome by electron diffraction. There is no problem of overlapping reflections in electron diffraction and all diffraction spots can be unambiguously indexed. There is no problem of under-determination (less data than unknown parameters) for electron diffraction since 10-100 times more reflections than parameters can be obtained by ED, whereas in X-ray powder diffraction the over-determination is close to one.

- HREM images can be used for studying defects, local structure and chemical variations in crystals, as well as interfaces.

Electron crystallography can be used to extend the range of samples amenable to structure analysis beyond those which can be studied by single crystal X-ray diffraction. Moreover, HREM images can supply some initial low resolution phases for X-ray diffraction which may aid in phase determination in X-ray crystallography.

Structure Refinement

A complete crystal structure determination (after collecting the data and finding the unit cell and the symmetry) can be divided in two distinct steps: solving the structure and refining it. Solving the structure means finding a rough model of at least the most important (= heaviest) atoms. For solving a structure, any method, including guessing, may be used.

The structure model arrived at from for example HREM images needs to be verified and further improved. This can be done by refinement of the structure model by least-squares methods against data from selected area electron diffraction or convergent beam electron diffraction. The accuracy of atomic coordinates obtained from a 2D projection by HREM to 2 Å resolution is about 0.2 Å. This limited accuracy is due only to the sparse amount of experimental data, with typically no more than one reflection per atomic coordinate. Electron diffraction data extends to much higher resolution, typically 0.8 to 0.5 Å, and so contains about ten times more reflections than the corresponding HREM image. These data have the further advantage compared to the image data that they are not distorted by the contrast transfer function (CTF) of the lens. The fact that SAED patterns do not contain phase information is no disadvantage for the refinement, because only the amplitudes are used at this stage of a structure determination.

Structure refinement based on kinematical scattering was already applied by the Russian scientist 60 years ago. Weirich et al. first solved the structure of an unknown $Ti11Se_4$ by HREM combined with crystallographic image processing. Then they used intensities extracted from selected area electron diffraction patterns of a very thin crystal and refined the structure to a precision of 0.02 Å for all the atoms. Wagner and Terasaki et al. determined the 3D structure of a new zeolite from selected area electron diffraction, based on kinematical approach.

The disadvantage selected area electron diffraction is that they contain a higher proportion of multiply scattered electrons than HREM images, mainly due to the factor that the SAED patterns are usually taken from a larger and thicker region than those thinnest parts of HREM images which are cut out and used for image processing.

Structure refinement based on dynamical scattering was developed by Zandbergen and Jansen, known as the MSLS software. Electron diffraction from crystal regions with relatively homogenous thicknesses was used. Both the crystal orientation, crystal thickness and the atomic coordinates could be refined simultaneously.

Electron Crystallography – The Future of Crystallography

Electron crystallography is still a young research field. The continuous development of modern electron microscopy techniques challenges scientists to explore the possibilities and push the limit of electron crystallography, and to tackle more difficult problems.

Extending Image Resolution by Through-focus Exit Wave Reconstruction

Modern electron microscopes with field emission electron sources provide brighter and more coherent electrons. Images with information of crystal structures up to 1 Å can be achieved. A through-focus exit wave reconstruction method was developed by Coene et al. to retrieve the complete exit wave function of electrons at the exit surface of the crystal. This method can be applied to thicker crystals which cannot be treated as weak-phase object. It is especially useful for studying defects and interfaces.

Electron Diffraction Imaging on Nanostructured Materials

Most of the phasing methods on electron diffraction, such as the direct methods and Patterson method, are only valid for crystalline materials with large number of repeating unit cell. When electron diffraction patterns are taken a finite sample (with the size only a few times of the unit cell dimensions) using a coherent electron beam, they can be sampled at a spacing finer than the inverse of the size of the sample (Nyquist frequency). In such a case, there will be a region of zeros (outside the sample) in the inverse Fourier transform of the diffraction pattern. The finer the sampling frequency, the larger the region of zeros. When the region of zeros is larger than the sample, the phase at each point of the whole electron diffraction pattern (both on and between the diffraction spots) could be directly retrieved using the iterative phase retrieval algorithm. This approach was applied to simulated electron diffraction patterns and they could show that 3D structure of a zeolite nanocrystal could be determined ab initio at a resolution of 1 Å from 29 electron diffraction patterns. This oversampling phasing method was demonstrated by Zuo et al. to imaging the 3D structure of a double carbon nanotube.

Lens-aberration Corrected Electron Microscopy

The point resolution of an electron microscope is limited by the spherical aberration of the objective lens. Haider et al. developed a corrector to be implemented into a standard transmission electron microscope to correct the spherical aberration. They showed that the point resolution could be improved from 2.4 Å to 1.4 Å.

With the help of the the spherical aberration corrector, Jia et al. could adjust the spherical aberration coefficient of the objective lens to a negative value and imaging directly oxygen atoms in perovskite ceramics $SrTiO_3$ and high-Tc superconductor YBa_2Cu_3O by HREM. They could see oxygen deficency and ordering directly from the HREM images. They could even measure the oxygen content at a twin boundary in $SrTiO_3$.

3D Reconstruction from HREM Images on Inorganic Crystals

HREM images are only projections of the 3D structure. Most of the structure determination from HREM images of inorganic compounds are on crystals with at least one short unit cell axis (<5 Å). The structures can be solved from a single projection with a short axis, where at least the heavy atoms do not overlap. Although the resolution of electron microscopes can be improved by the spherical aberration correction, for complex structures with large unit cell axes, atoms may still overlap in HREM images even at a resolution of better than 1Å. The only way to solve the overlapping problem is to collect several images from different directions and combine these images into a 3D structure. Nowadays electron microscopes can provide images with resolutions beyond 1.5 Å for inorganic crystals. This resolution is sufficient to resolve interatomic spacing, which means that it should be possible to resolve all atoms.

In 1992 Wenk et al. in Berkeley for the first time combined 2D HREM images into a 3D reconstruction for solving the structure of an inorganic crystal. They combined, using image processing, HREM images (to a resolution of 1.38 Å) taken in five different orientations of the silicate mineral staurolite and constructed a 3D electron potential map. In this map all atoms (Fe, Al, Si and oxygen) were clearly resolved. This work showed that 3D electron crystallography has a great potential also in structure determination of inorganic crystals - perhaps even more promising for inorganic structures than for organic and biological structures, because of the higher resolution.

We have reconstructed the 3D structure of a complex quasicrystal approximant ν-AlCrFe ($P6_3/m$, a = 40.687 and c = 12.546 Å). Due to the huge unit cell, it was necessary to combine crystallographic data from 13 projections to resolve the atoms. Electron microscopy images containing both amplitude and phase information were combined with amplitudes from electron diffraction patterns. 124 of the 129 unique atoms (1176 in the unit cell) were found in the remarkably clean calculated potential maps. This investigation demonstrates that inorganic crystals of any complexity can be solved by electron crystallography.

The 3D reconstruction method was also applied to solve the 3D structures of a series mesoporous material.

Ultrafast Electron Crystallography

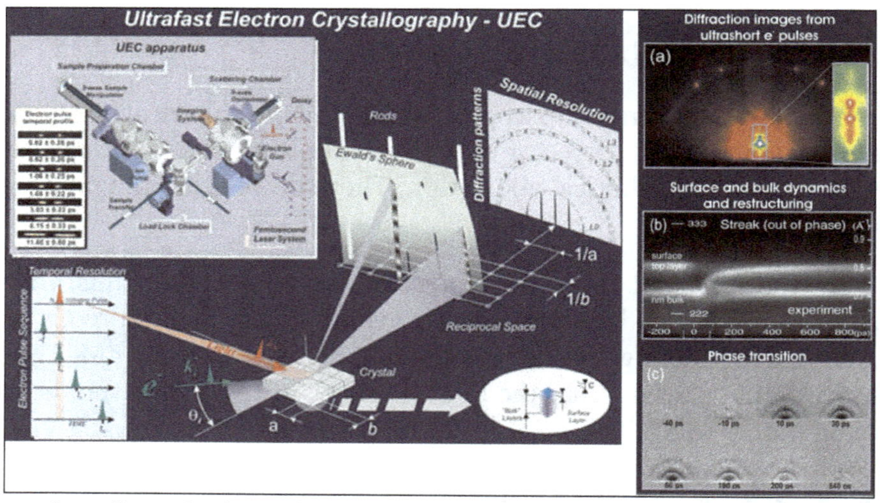

Ultra-fast Electron Crystallography for studying transient structures of molecules, surfaces and phase transitions.

Most of chemical reactions and phase transitions occur in a time scale of femto- to pico- seconds. Electron crystallography provides the only technique to determine the entire structure changes within such a time scale, thanks for the strong interaction between electrons and matter. Recently femtosecond ultrafast electron crystallography was developed by Zewail's group and Siwick et al. Zewail was awarded the Nobel Prize in 1999 for his studies of the transition states of chemical reactions using femtosecond spectroscopy. The ultrafast electron diffraction technique employs properly timed sequences of ultrafast pulses - a femtosecond laser pulse - to initiate the reaction and then followed by an ultrashort electron pulses to probe the ensuing structural change in the sample. The resulting electron diffraction patterns are then recorded on a CCD camera. This sequence of pulses is repeated, timing the electron pulse to arrive before or after the laser pulse; in effect, a series of snapshots of the evolving molecular structure are taken. Each time-resolved diffraction pattern can then, in principle, be inverted to reveal the three-dimensional molecular structure that gave rise to the pattern at that specific time delay. However, in practice, a key challenge lies in recovering the molecular structural information that is embedded in the as-acquired diffraction images. Another challenge is, to obtain EM images of the sample and retrieve the structure information from them.

Electron crystallography provides unique possibilities of studying nanostructured materials. The aims of this course are to show theoretically that electron crystallography can be used for crystal structure analysis; to describe different methods of solving crystal structures by electron crystallography and to demonstrate how these methods are used practically.

Optical Crystallography

Optical crystallography is branch of crystallography that deals with the optical properties of crystals. It is of considerable interest theoretically and has the greatest practical importance. The science of petrography is largely based on the study of the appearance of thin, transparent sections of rocks in a microscope fitted with polarizers. In the absence of external crystalline form, as with the minerals in a rock, a mineral often may be readily identified by the determination of some of its optical properties.

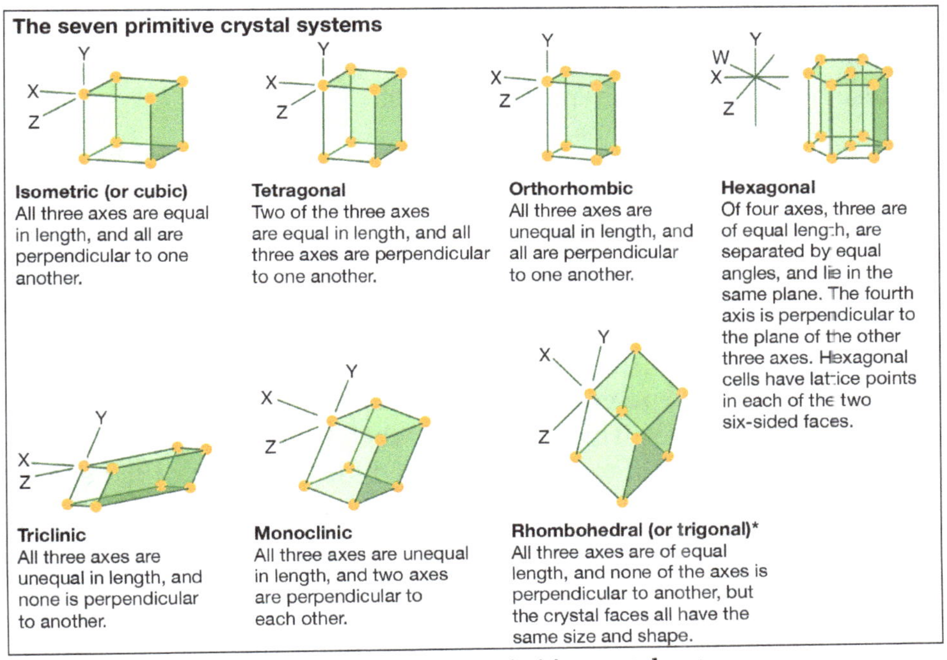

Crystal systems: The seven primitive crystal systems.

According to their action on transmitted plane-polarized light, all crystals may be assigned to one or another of the five groups enumerated below, which correspond to the seven systems of crystallization (in the second group, three systems are included together). The several symmetry classes of each system are optically the same, except in rare cases:

- Optically isotropic crystals (isometric system), which exhibit only one index of refraction for light of each colour;
- Optically uniaxial crystals (tetragonal, hexagonal, and rhombohedral [or trigonal] systems), which exhibit double refraction and yield two refractive indices for light of each colour, one parallel to the optical axis and one perpendicular to the optical axis;
- Optically biaxial crystals (all of which exhibit three principal refractive indices, one along each of the mutually perpendicular optical axes) in which the three optical axes correspond to the three crystallographic axes (orthorhombic system);
- Optically biaxial crystals in which only one of the three optical axes corresponds to a crystallographic axis (monoclinic system);
- Optically biaxial crystals with no fixed and definite relation between the optical and crystallographic axes (triclinic system).

The study of the propagation of light, and associated phenomena, in crystalline solids. For a simple cubic crystal the atomic arrangement is such that in each direction through the crystal the crystal presents the same optical appearance. The atoms in anisotropic crystals are closer together in some planes through the material than in others. In anisotropic crystals the optical characteristics are different in different directions. In classical physics the progress of an electromagnetic wave through a material involves the periodic displacement of electrons. In anisotropic substances the forces resisting these displacements depend on the displacement direction. Thus the velocity of a light wave is different in different directions and for different states of polarization. The absorption of the wave may also be different in different directions.

In an isotropic medium the light from a point source spreads out in a spherical shell. The light from a point source embedded in an anisotropic crystal spreads out in two wave surfaces, one of which travels at a faster rate than the other. The polarization of the light varies from point to point over each wave surface, and in any particular direction from the source the polarization of the two surfaces is opposite. The characteristics of these surfaces can be determined experimentally by making measurements on a given crystal.

In the most general case of a transparent anisotropic medium, the dielectric constant is different along each of three orthogonal axes. This means that when the light vector is oriented along each direction, the velocity of light is different. One method for calculating the behavior of a transparent anisotropic material is through the use of the index ellipsoid, also called the reciprocal ellipsoid, optical indicatrix, or ellipsoid of wave normals. This is the surface obtained by plotting the value of the refractive index in each principal direction for a linearly polarized light vector lying in that direction. The different indices of refraction, or wave velocities associated with a given propagation direction, are then given by sections through the origin of the coordinates in which the index ellipsoid is drawn. These sections are ellipses, and the major and minor axes of the ellipse represent the fast and slow axes for light proceeding along the normal to the plane of the ellipse. The length of the axes represents the refractive indices for the fast and slow wave, respectively. The most asymmetric type of ellipsoid has three unequal axes. It is a general rule in crystallography that no property of a crystal will have less symmetry than the class in which the crystal belongs.

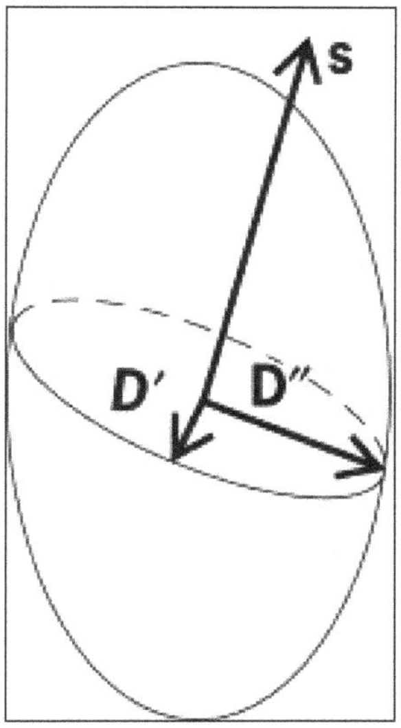

Index ellipsoid, showing construction of directions of vibrations of D vectors belonging to a wave normal s.

Accordingly, there are many crystals which, for example, have four- or sixfold rotation symmetry about an axis, and for these the index ellipsoid cannot have three unequal axes but is an ellipsoid of revolution. In such a crystal, light will be propagated along this axis as though the crystals were isotropic, and the velocity of propagation will be independent of the state of polarization. The section of the index ellipsoid at right angles to this direction is a circle. Such crystals are called uniaxial and the mathematics of their optical behavior is relatively straightforward.

In crystals of low symmetry the index ellipsoid has three unequal axes. These crystals are termed biaxial and have two directions along which the wave velocity is independent of the polarization direction. These correspond to the two sections of the ellipsoid which are circular.

The normal to a plane wavefront moves with the phase velocity. The Huygens wavelet, which is the light moving out from a point disturbance, will propagate with a ray velocity. Just as the index ellipsoid can be used to compute the phase or wave velocity, so can a ray ellipsoid be used to calculate the ray velocity. The length of the axes of this ellipsoid is given by the velocity of the linearly polarized ray whose electric vector lies in the axis direction.

The refraction of a light ray on passing through the surface of an anisotropic uniaxial crystal can be calculated with Huygens wavelets in the same manner as in an isotropic material. For the ellipsoidal wavelet this results in an optical behavior which is completely different from that normally associated with refraction. The ray associated with this behavior is termed the extraordinary ray. At a crystal surface where the optic axis is inclined at an angle, a ray of unpolarized light incident normally on the surface is split into two beams: the ordinary ray, which proceeds through the surface without deviation; and the extraordinary ray, which is deviated by an angle determined by a line drawn from the center of one of the Huygens ellipsoidal wavelets to the point at which the ellipsoid is tangent to a line parallel to the surface. The two beams are oppositely linearly polarized.

Crystal Optics an area between optics and crystal physics, encompassing the study of the propagation of light in crystals. Phenomena characteristic of crystals that are studied in crystal optics include double refraction, polarization of light, rotation of the plane of polarization, and pleochroism.

The phenomenon of double refraction was first observed in crystals of Iceland spar by the Danish scientist E. Bartholin in 1669. This date is considered the beginning of crystal optics. Problems of the absorption and emission of light by crystals are studied in crystal spectroscopy. The effect of electric and magnetic fields on the optical properties of crystals is studied in electro-optics and magneto-optics, which are based on the fundamental laws of crystal optics.

Since the lattice constant (of the order of 10 angstroms [Å]) is much smaller than the wavelength of visible light (4,000–7,000 Å), a crystal may be regarded as a homogeneous but anisotropic medium. The optical anisotropy of crystals is caused by the anisotropy of the force field of particle interaction. The nature of the field is related to crystal symmetry. All crystals, except crystals of the cubic system, are optically anisotropic.

The optical anisotropy of transparent nonmagnetic crystals is due to the anisotropy of the dielectric constant c. In isotropic mediums the electric induction vector D is related to the electric field vector E by the equation $D = \epsilon E$, where ϵ is a scalar quantity that depends on the frequency in the case of alternating fields. Thus, the vectors D and E have the same direction in isotropic mediums. In crystals the directions of the vectors D and E do not coincide, and the relationship between D and E is more complex, since the dielectric constant e, a tensor quantity, depends on direction in the crystal. A result of this is the observed anisotropy of the optical properties of crystals, particularly the dependence on direction of the rate of propagation v of waves and of the index of refraction n. The dependence of the components of the dielectric constant tensor on wave frequency explains the dispersion of the optical properties of crystals.

The dependence on direction of the dielectric constant ϵ—and, consequently, of the index of refraction n—can be represented graphically. If radius vectors r with magnitudes $r = n = \sqrt{\epsilon}$, where ϵ is the dielectric constant in direction r, are drawn in all directions from an arbitrary point O of the crystal, then the ends of the vectors r lie on the surface of an ellipsoid, called the optical indicatrix or index ellipsoid. The axes of symmetry of the ellipsoid define the three mutually perpendicular principal directions in the crystal. In a rectangular Cartesian coordinate system, whose axes coincide with the principal directions, the equation of the optical indicatrix is:

$$\frac{x^2}{n_{x^2}}+\frac{y^2}{n_{y^2}}+\frac{z^2}{n_{z^2}}=1$$

Where, n_x, n_y, and n_z are the values of n along the principal directions (principal values of the tensor of the dielectric constant and of n). The optical axis of the crystal is a straight line that passes through a given point O of the crystal at right angles to the plane of a circular section of the optical indicatrix.

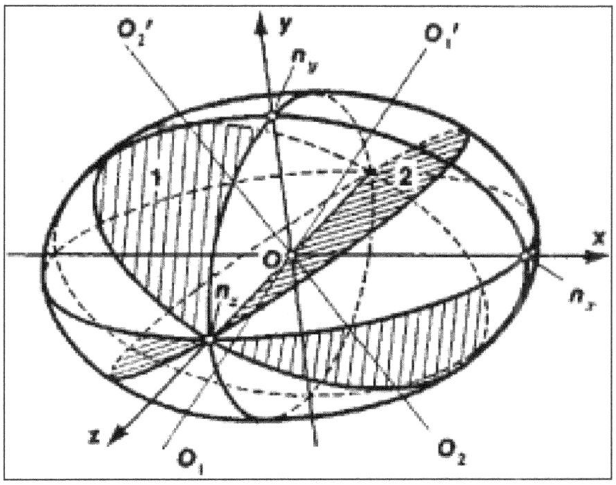

Optical indicatrix of a biaxial crystal, a triaxial ellipsoid: (o_x), (o_y), and (o_z) axes of symmetry of the ellipsoid, called the principal axes of the indicatrix; (n_x), (n_y), and (n_z) indexes of refraction along the principal axes; (1) and (2) two circular sections of the ellipsoid; ($O_1O'_1$) and ($O_2O'_2$) optical axes of the crystal.

In the case of cubic crystals, which are optically isotropic, ∊ is independent of direction, and the optical indicatrix becomes a sphere with radius $r = n = \sqrt{\epsilon}$. In crystals of intermediate systems (trigonal, tetragonal, and hexagonal), one of the principal directions coincides with the main axis of symmetry of the crystal. The optical indicatrix in such crystals is an ellipsoid of revolution, and the crystals have only one optical axis, which coincides with the axis of rotation of the ellipsoid. Such crystals are called uniaxial. A uniaxial crystal is called optically positive (+) if it's optical axis coincides with the long axis of the optical indicatrix (the ellipsoid is elongated along the axis of rotation) and optically negative (–) if the ellipsoid is compressed along the axis of rotation. Crystals of the lower systems (rhombic, monoclinic, and triclinic) are called biaxial. Their optical indicatrix is a triaxial ellipsoid, which has two circular sections and two optical axes.

Because of the noncoincidence of the direction of the vectors D and E, a polarized, plane monochromatic wave in a crystal is characterized by two triads of mutually perpendicular vectors D, H, v and E, H, v′. The velocity v′ coincides in direction with the Poynting vector S and is equal to the rate of energy transfer by the wave. It is called the ray velocity of the wave. The velocity v is called the normal velocity of the wave. It is equal to the rate of propagation of the phase and the wave front in the direction of the normal vector N to the front. The quantities v and v′ are connected by the relation $v' = v/\cos\alpha$, where α is the angle between the vectors D and E.

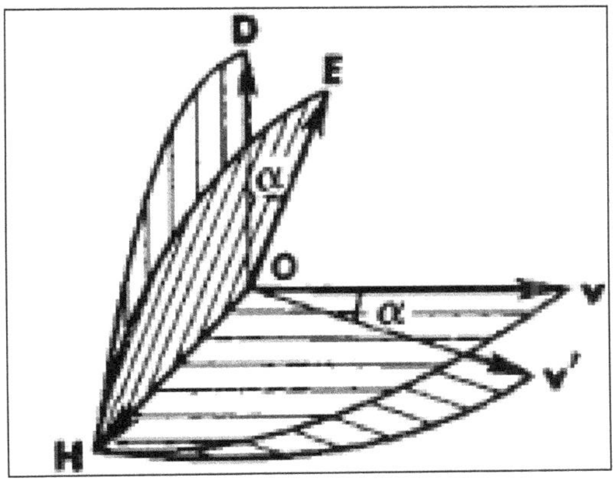

The normal and ray velocities of the wave are determined from the Fresnel equation, which is the basic equation of crystal optics:

$$\frac{N_{x^2}}{v^2 - v_{x^2}} + \frac{N_{y^2}}{v^2 - v_{y^2}} + \frac{N_{z^2}}{v^2 - v_{z^2}} = 0$$

Here N_x, N_y, and N_z are projections of the vector normal N onto the principal directions of the crystal; $v_x = c/n_y$, $v_y = c/n_z$ are the principal phase velocities of the wave; c is the velocity of light in a vacuum; and n_x, n_y, n_z are the principal indexes of refraction of the crystal.

Since the Fresnel equation is quadratic with respect to v, for any direction N there are two values of the normal velocity, v_1 and v_2, which coincide only in the direction of the optical axes of the crystal. If vectors of the corresponding normal velocities v are plotted from a point O in all directions N, then the ends of the vectors lie on a surface called the normal surface. It is a double-sheeted surface, each sheet of which corresponds to one of the values of v for a given direction N. In the case of a uniaxial crystal, one surface is a sphere and the other is an ovaloid, which touches the sphere at its two points of intersection with the optical axis. In the case of a biaxial crystal, the surfaces intersect at four points located on the two optical axes (binormals).

Similarly, the geometric locus of points at a distance v′ from a point O is called the ray surface, or wave surface. It is the wave surface for waves propagating within the crystal from a point source located at point O. This is also a double-sheeted surface. In uniaxial crystals, one surface is a sphere and the other is an ellipsoid of revolution about the optical axis Oz. The sphere and the ellipsoid touch in the points of their intersection with the optical axis. In positive crystals the ellipsoid is inscribed within the sphere, whereas in negative crystals the sphere is inscribed within the ellipsoid. In biaxial crystals, the surfaces intersect at four points located in pairs on two straight lines that intersect at point O (biradials).

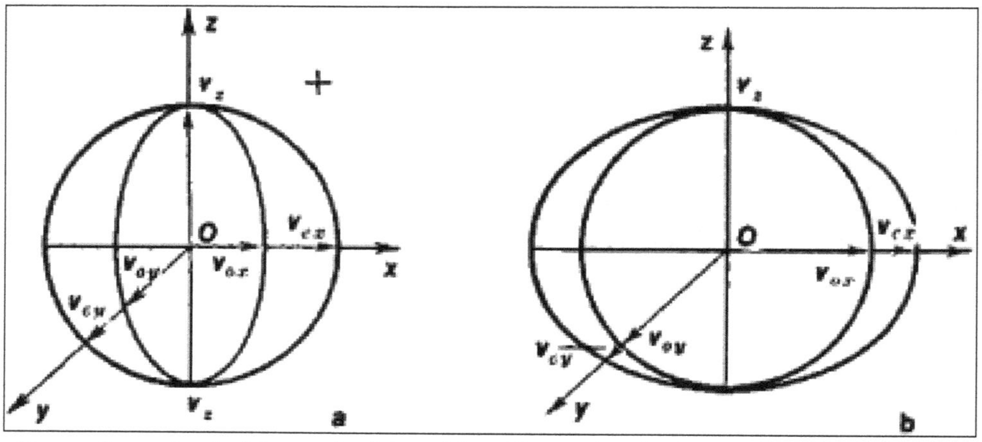

Ray surfaces of uniaxial crystals: (a) positive, (b) negative, (OZ) optical axis of the crystal, (v_o) and (v_c) phase velocities of two waves propagating in the crystals.

Thus, two plane waves that are polarized in two mutually perpendicular planes may propagate in an arbitrary direction N in crystals. The directions of the vectors D_1 and D_2 of the waves coincide with the axes of the ellipse resulting from intersection of the optical indicatrix with the plane perpendicular to N and passing through the point O. The normal velocities of the waves are $v_1 = c/n_1$

and $v_2 = c/n_2$. The vectors E_1 and E_2 of the waves are also located on two perpendicular planes, and there are two corresponding ray vectors, Si and S2, and two values of the ray velocities, $v_1 = v_1' \cos \alpha$ and $v_2 = v_2' \cos \alpha$. Similarly, for a given direction of the ray S there are two possible directions for the vector E ($E_1 \perp E_2$), corresponding to the two values of the ray velocity, v_1' and v_2'.

The dependence of the ray velocity of a plane wave propagating in a crystal on the direction of propagation and the nature of polarization of the wave leads to the splitting of light rays in crystals. In a uniaxial crystal, one of the refracted rays obeys the usual laws of refraction and is therefore called the ordinary ray, whereas the other ray does not (it does not lie in the plane of incidence) and is called the extraordinary ray. In a biaxial crystal, both rays are extraordinary.

During propagation in a crystal, the two waves arising upon refraction acquire a path difference because of differences in the indexes of refraction and the geometric path, although they re-main coherent. A polarization device may be used to bring the directions of oscillation of the waves emerging from the crystal into the same plane and to observe their interference. Interference of linearly polarized white light is manifested as coloration of the crystal, which depends on the path difference acquired by the beam. Characteristic interference patterns ("conoscopic patterns") whose shape depends on the orientation of the crystal may sometimes be observed.

In crystals of some symmetry classes, rotation of the plane of polarization is possible in addition to double refraction. Two elliptically polarized waves (with opposite senses of rotation), each with its own index of refraction, may propagate in each direction in such crystals. The polarization is circular only in the direction of the optical axis, which leads to rotation of the plane of polarization of linearly polarized light incident on the crystal.

In the case of strongly absorbing crystals, a linearly polarized wave propagating in the crystal is split into two elliptically polarized waves with the same sense of rotation. Such crystals exhibit different absorption of waves with different polarization.

Each crystal has its own characteristic set of optical proper-ties, which can be used for identification. The most important properties for uniaxial crystals are the indexes of refraction n_o of the ordinary wave and n_e of the extraordinary wave and the difference Δn between them (magnitude of the double refraction), as well as the dependence of the characteristics on the wave-length (various types of dispersion). Biaxial crystals are characterized by a more complex set of properties. In applied crystal optics, which is concerned with analysis of minerals and rocks, various methods for measuring these quantities were developed for various preparations of minerals in the form of powders and thin plates (sections). The principal methods are the immersion method, for determining the index of refraction using special liquids or melts with known indexes of refraction, and Fedorov's method, for the determination of the orientation of the indicatrix, using a microscope stage that rotates the crystal about various axes. Most crystal-optics measurements are performed with a polarization microscope.

4

Crystal Structure and Crystal Systems

A crystal structure is defined as the ordered arrangements of atoms, ions and molecules in a crystal. The space groups, lattices, point groups, crystal forms, etc. together form a crystal system. This chapter closely examines these concepts of crystal structure and crystal system to provide an extensive understanding of the subject.

Crystal Structure

The crystal structure is formed by associating every lattice point with an assembly of atoms or molecules or ions, which are identical in composition, arrangement and orientation, is called as the basis. The atomic arrangement in a crystal is called crystal structure. If the basis is substituted for the lattice points, then the resulting structure is called crystal structure. Thus lattice + basis = crystal structure. In copper and sodium crystals the basis is single atoms; in NaCl, the basis is diatomic and in CaF_2 the basis is triatomic.

The smallest group of particles in the material that constitutes this repeating pattern is the unit cell of the structure. The unit cell completely reflects the symmetry and structure of the entire crystal, which is built up by repetitive translation of the unit cell along its principal axes. The translation vectors define the nodes of the Bravais lattice.

The lengths of the principal axes, or edges, of the unit cell and the angles between them are the lattice constants, also called lattice parameters or cell parameters. The symmetry properties of the crystal are described by the concept of space groups. All possible symmetric arrangements of particles in three-dimensional space may be described by the 230 space groups.

The crystal structure and symmetry play a critical role in determining many physical properties, such as cleavage, electronic band structure, and optical transparency.

Unit Cell

Crystal structure is described in terms of the geometry of arrangement of particles in the unit cell. The unit cell is defined as the smallest repeating unit having the full symmetry of the crystal structure. The geometry of the unit cell is defined as a parallelepiped, providing six lattice parameters taken as the lengths of the cell edges (a, b, c) and the angles between them (α, β, γ). The positions of particles inside the unit cell are described by the fractional coordinates (x_i, y_i, z_i) along the cell edges, measured from a reference point. It is only necessary to report the coordinates of a smallest asymmetric subset of particles. This group of particles may be chosen so that it occupies the smallest physical space, which means that not all particles need to be physically located inside the boundaries given by the lattice parameters. All other particles of the unit cell are generated by the symmetry operations that characterize the symmetry of the unit cell. The collection of symmetry operations of the unit cell is expressed formally as the space group of the crystal structure.

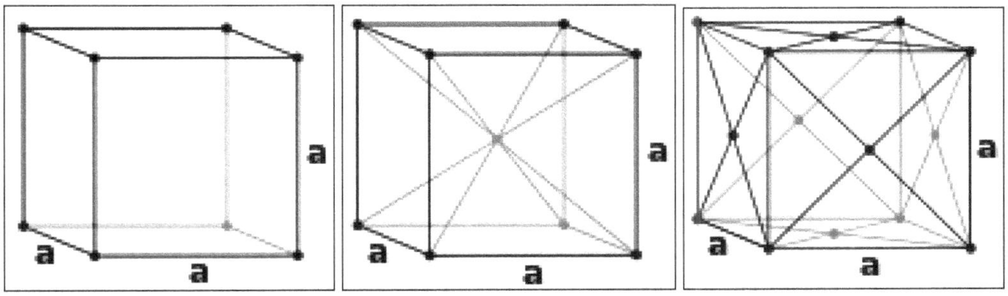

Simple cubic (P) Body-centered cubic (I) Face-centered cubic (F).

Miller Indices

Vectors and planes in a crystal lattice are described by the three-value Miller index notation. This syntax uses the indices ℓ, *m*, and *n* as directional parameters.

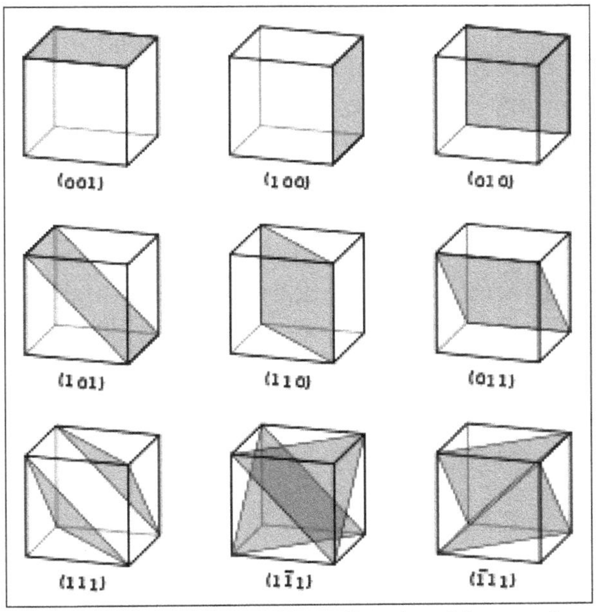

Planes with different Miller indices in cubic crystals.

By definition, the syntax (ℓ*mn*) denotes a plane that intercepts the three points a_1/ℓ, a_2/m, and a_3/n, or some multiple thereof. That is, the Miller indices are proportional to the inverses of the intercepts of the plane with the unit cell (in the basis of the lattice vectors). If one or more of the indices is zero, it means that the planes do not intersect that axis (i.e., the intercept is "at infinity"). A plane containing a coordinate axis is translated so that it no longer contains that axis before its Miller indices are determined. The Miller indices for a plane are integers with no common factors. Negative indices are indicated with horizontal bars, as in $(\bar{1}23)$. In an orthogonal coordinate system for a cubic cell, the Miller indices of a plane are the Cartesian components of a vector normal to the plane.

Considering only (ℓmn) planes intersecting one or more lattice points (the lattice planes), the distance d between adjacent lattice planes is related to the (shortest) reciprocal lattice vector orthogonal to the planes by the formula:

$$d = \frac{2\pi}{|g_{\ell mn}|}$$

Planes and Directions

The crystallographic directions are geometric lines linking nodes (atoms, ions or molecules) of a crystal. Likewise, the crystallographic planes are geometric planes linking nodes. Some directions and planes have a higher density of nodes. These high density planes have an influence on the behavior of the crystal as follows:

- Optical properties: Refractive index is directly related to density (or periodic density fluctuations).
- Adsorption and reactivity: Physical adsorption and chemical reactions occur at or near surface atoms or molecules. These phenomena are thus sensitive to the density of nodes.
- Surface tension: The condensation of a material means that the atoms, ions or molecules are more stable if they are surrounded by other similar species. The surface tension of an interface thus varies according to the density on the surface.
- Microstructural defects: Pores and crystallites tend to have straight grain boundaries following higher density planes.
- Cleavage: This typically occurs preferentially parallel to higher density planes.
- Plastic deformation: Dislocation glide occurs preferentially parallel to higher density planes. The perturbation carried by the dislocation (Burgers vector) is along a dense direction. The shift of one node in a more dense direction requires a lesser distortion of the crystal lattice.

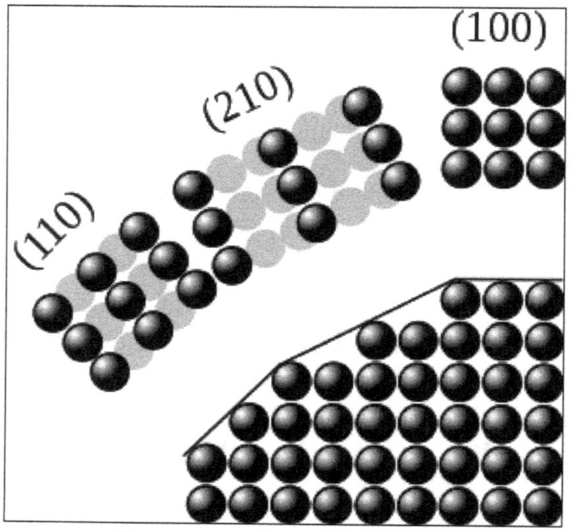

Dense crystallographic planes.

Some directions and planes are defined by symmetry of the crystal system. In monoclinic, rhombohedral, tetragonal, and trigonal/hexagonal systems there is one unique axis (sometimes called the principal axis) which has higher rotational symmetry than the other two axes. The basal plane is the plane perpendicular to the principal axis in these crystal systems. For triclinic, orthorhombic, and cubic crystal systems the axis designation is arbitrary and there is no principal axis.

Cubic Structures

For the special case of simple cubic crystals, the lattice vectors are orthogonal and of equal length (usually denoted *a*); similarly for the reciprocal lattice. So, in this common case, the Miller indices (*ℓmn*) and [*ℓmn*] both simply denote normals/directions in Cartesian coordinates. For cubic crystals with lattice constant *a*, the spacing *d* between adjacent (ℓmn) lattice planes is:

$$d_{\ell mn} = \frac{a}{\sqrt{\ell^2 + m^2 + n^2}}$$

Because of the symmetry of cubic crystals, it is possible to change the place and sign of the integers and have equivalent directions and planes:

- Coordinates in angle brackets such as ⟨100⟩ denote a family of directions that are equivalent due to symmetry operations, such as,, or the negative of any of those directions.
- Coordinates in curly brackets or braces such as {100} denote a family of plane normals that are equivalent due to symmetry operations, much the way angle brackets denote a family of directions.

For face-centered cubic (fcc) and body-centered cubic (bcc) lattices, the primitive lattice vectors are not orthogonal. However, in these cases the Miller indices are conventionally defined relative to the lattice vectors of the cubic supercell and hence are again simply the Cartesian directions.

Interplanar Spacing

The spacing d between adjacent (*hkl*) lattice planes is given by:

- Cubic:

$$\frac{1}{d^2} = \frac{h^2 + k^2 + l^2}{a^2}$$

- Tetragonal:

$$\frac{1}{d^2} = \frac{h^2 + k^2}{a^2} + \frac{l^2}{c^2}$$

- Hexagonal:

$$\frac{1}{d^2} = \frac{4}{3}\left(\frac{h^2 + hk + k^2}{a^2}\right) + \frac{l^2}{c^2}$$

- Rhombohedral:

$$\frac{1}{d^2} = \frac{(h^2+k^2+l^2)\sin^2\alpha + 2(hk+kl+hl)(\cos^2\alpha - \cos\alpha)}{a^2(1-3\cos^2\alpha+2\cos^3\alpha)}$$

- Orthorhombic:

$$\frac{1}{d^2} = \frac{h^2}{a^2} + \frac{k^2}{b^2} + \frac{l^2}{c^2}$$

- Monoclinic:

$$\frac{1}{d^2} = \left(\frac{h^2}{a^2} + \frac{k^2\sin^2\beta}{b^2} + \frac{l^2}{c^2} - \frac{2hl\cos\beta}{ac}\right)\csc^2\beta$$

- Triclinic:

$$\frac{1}{d^2} = \frac{\frac{h^2}{a^2}\sin^2\alpha + \frac{k^2}{b^2}\sin^2\beta + \frac{l^2}{c^2}\sin^2\gamma + \frac{2kl}{bc}(\cos\beta\cos\gamma - \cos\alpha) +}{1-\cos^2\alpha-\cos^2\beta-\cos^2\gamma+2\cos\alpha\cos\beta\cos\gamma} \cdots$$

$$\frac{\frac{2hl}{ac}(\cos\gamma\cos\alpha - \cos\beta) + \frac{2hk}{ab}(\cos\alpha\cos\beta - \cos\gamma)}{1-\cos^2\alpha-\cos^2\beta-\cos^2\gamma+2\cos\alpha\cos\beta\cos\gamma}$$

Atomic Coordination

By considering the arrangement of atoms relative to each other, their coordination numbers (or number of nearest neighbors), interatomic distances, types of bonding, etc., it is possible to form a general view of the structures and alternative ways of visualizing them.

Close Packing

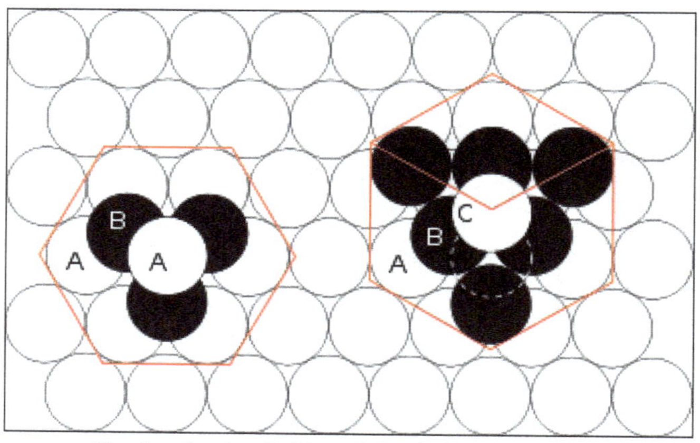

The hcp lattice (left) and the fcc lattice (right).

The principles involved can be understood by considering the most efficient way of packing together equal-sized spheres and stacking close-packed atomic planes in three dimensions. For example, if plane A lies beneath plane B, there are two possible ways of placing an additional atom on top of layer B. If an additional layer was placed directly over plane A, this would give rise to the following series:

...ABABABAB...

This arrangement of atoms in a crystal structure is known as hexagonal close packing (hcp).

If, however, all three planes are staggered relative to each other and it is not until the fourth layer is positioned directly over plane A that the sequence is repeated, then the following sequence arises:

...ABCABCABC...

This type of structural arrangement is known as cubic close packing (ccp).

The unit cell of a ccp arrangement of atoms is the face-centered cubic (fcc) unit cell. This is not immediately obvious as the closely packed layers are parallel to the {111} planes of the fcc unit cell. There are four different orientations of the close-packed layers.

The packing efficiency can be worked out by calculating the total volume of the spheres and dividing by the volume of the cell as follows:

$$\frac{4\times\frac{4}{3}\pi r^3}{16\sqrt{2}r^3}=\frac{\pi}{3\sqrt{2}}=0.7405...$$

The 74% packing efficiency is the maximum density possible in unit cells constructed of spheres of only one size. Most crystalline forms of metallic elements are hcp, fcc, or bcc (body-centered cubic). The coordination number of atoms in hcp and fcc structures is 12 and its atomic packing factor (APF) is the number mentioned above, 0.74. This can be compared to the APF of a bcc structure, which is 0.68.

Grain Boundaries

Grain boundaries are interfaces where crystals of different orientations meet. A grain boundary is a single-phase interface, with crystals on each side of the boundary being identical except in orientation. The term "crystallite boundary" is sometimes, though rarely, used. Grain boundary areas contain those atoms that have been perturbed from their original lattice sites, dislocations, and impurities that have migrated to the lower energy grain boundary.

Treating a grain boundary geometrically as an interface of a single crystal cut into two parts, one of which is rotated, we see that there are five variables required to define a grain boundary. The first two numbers come from the unit vector that specifies a rotation axis. The third number designates the angle of rotation of the grain. The final two numbers specify the plane of the grain boundary (or a unit vector that is normal to this plane).

Grain boundaries disrupt the motion of dislocations through a material, so reducing crystallite size is a common way to improve strength, as described by the Hall–Petch relationship. Since grain boundaries are defects in the crystal structure they tend to decrease the electrical and thermal conductivity of the material. The high interfacial energy and relatively weak bonding in most grain boundaries often makes them preferred sites for the onset of corrosion and for the precipitation of new phases from the solid. They are also important to many of the mechanisms of creep.

Grain boundaries are in general only a few nanometers wide. In common materials, crystallites are large enough that grain boundaries account for a small fraction of the material. However, very small grain sizes are achievable. In nanocrystalline solids, grain boundaries become a significant volume fraction of the material, with profound effects on such properties as diffusion and plasticity. In the limit of small crystallites, as the volume fraction of grain boundaries approaches 100%, the material ceases to have any crystalline character, and thus becomes an amorphous solid.

Prediction of Structure

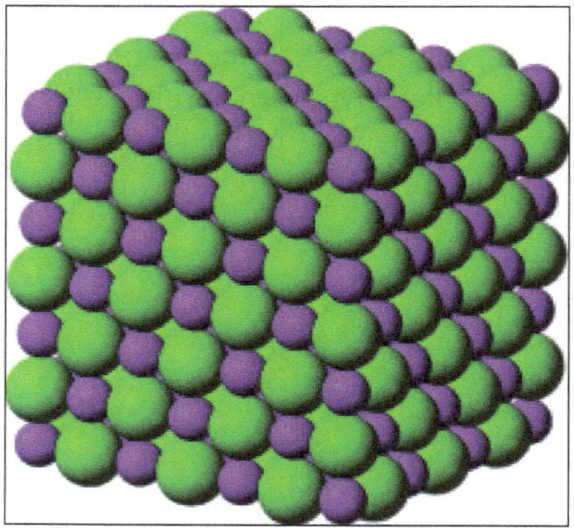

Crystal structure of sodium chloride (table salt).

The difficulty of predicting stable crystal structures based on the knowledge of only the chemical composition has long been a stumbling block on the way to fully computational materials design. Now, with more powerful algorithms and high-performance computing, structures of medium complexity can be predicted using such approaches as evolutionary algorithms, random sampling, or metadynamics.

The crystal structures of simple ionic solids (e.g., NaCl or table salt) have long been rationalized in terms of Pauling's rules, first set out in 1929 by Linus Pauling, referred to by many since as the "father of the chemical bond". Pauling also considered the nature of the interatomic forces in metals, and concluded that about half of the five d-orbitals in the transition metals are involved in bonding, with the remaining nonbonding d-orbitals being responsible for the magnetic properties. He, therefore, was able to correlate the number of d-orbitals in bond formation with the bond length as well as many of the physical properties of the substance. He subsequently introduced the metallic orbital, an extra orbital necessary to permit uninhibited resonance of valence bonds among various electronic structures.

In the resonating valence bond theory, the factors that determine the choice of one from among alternative crystal structures of a metal or intermetallic compound revolve around the energy of resonance of bonds among interatomic positions. It is clear that some modes of resonance would make larger contributions (be more mechanically stable than others), and that in particular a simple ratio of number of bonds to number of positions would be exceptional. The resulting principle is that a special stability is associated with the simplest ratios or "bond numbers": ½, ⅓, ⅔, ¼, ¾, etc. The choice of structure and the value of the axial ratio (which determines the relative bond lengths) are thus a result of the effort of an atom to use its valency in the formation of stable bonds with simple fractional bond numbers.

After postulating a direct correlation between electron concentration and crystal structure in beta-phase alloys, Hume-Rothery analyzed the trends in melting points, compressibilities and bond lengths as a function of group number in the periodic table in order to establish a system of valencies of the transition elements in the metallic state. This treatment thus emphasized the increasing bond strength as a function of group number. The operation of directional forces were emphasized in one article on the relation between bond hybrids and the metallic structures. The resulting correlation between electronic and crystalline structures is summarized by a single parameter, the weight of the d-electrons per hybridized metallic orbital. The "d-weight" calculates out to 0.5, 0.7 and 0.9 for the fcc, hcp and bcc structures respectively. The relationship between d-electrons and crystal structure thus becomes apparent.

In crystal structure predictions/simulations, the periodicity is usually applied, since the system is imagined as unlimited big in all directions. Starting from a triclinic structure with no further symmetry property assumed, the system may be driven to show some additional symmetry properties by applying Newton's Second Law on particles in the unit cell and a recently developed dynamical equation for the system period vectors (lattice parameters including angles), even if the system is subject to external stress.

Polymorphism

Quartz is one of the several crystalline forms of silica, SiO_2. The most important forms of silica include: α-quartz, β-quartz, tridymite, cristobalite, coesite, and stishovite.

Polymorphism is the occurrence of multiple crystalline forms of a material. It is found in many crystalline materials including polymers, minerals, and metals. According to Gibbs' rules of phase equilibria, these unique crystalline phases are dependent on intensive variables such as pressure and temperature. Polymorphism is related to allotropy, which refers to elemental solids. The complete morphology of a material is described by polymorphism and other variables such as crystal habit, amorphous fraction or crystallographic defects. Polymorphs have different stabilities and may spontaneously and irreversibly transform from a metastable form (or thermodynamically unstable form) to the stable form at a particular temperature. They also exhibit different melting points, solubilities, and X-ray diffraction patterns.

One good example of this is the quartz form of silicon dioxide, or SiO_2. In the vast majority of silicates, the Si atom shows tetrahedral coordination by 4 oxygens. All but one of the crystalline forms involves tetrahedral $\{SiO_4\}$ units linked together by shared vertices in different arrangements. In different minerals the tetrahedra show different degrees of networking and polymerization. For example, they occur singly, joined together in pairs, in larger finite clusters including rings, in chains, double chains, sheets, and three-dimensional frameworks. The minerals are classified into groups based on these structures. In each of the 7 thermodynamically stable crystalline forms or polymorphs of crystalline quartz, only 2 out of 4 of each the edges of the $\{SiO_4\}$ tetrahedra are shared with others, yielding the net chemical formula for silica: SiO_2.

Another example is elemental tin (Sn), which is malleable near ambient temperatures but is brittle when cooled. This change in mechanical properties due to existence of its two major allotropes, α- and β-tin. The two allotropes that are encountered at normal pressure and temperature, α-tin and β-tin, are more commonly known as *gray tin* and *white tin* respectively. Two more allotropes, γ and σ, exist at temperatures above 161 °C and pressures above several GPa. White tin is metallic, and is the stable crystalline form at or above room temperature. Below 13.2 °C, tin exists in the gray form, which has a diamond cubic crystal structure, similar to diamond, silicon or germanium. Gray tin has no metallic properties at all, is a dull gray powdery material, and has few uses, other than a few specialized semiconductor applications. Although the α–β transformation temperature of tin is nominally 13.2 °C, impurities (e.g. Al, Zn, etc.) lower the transition temperature well below 0 °C, and upon addition of Sb or Bi the transformation may not occur at all.

Physical Properties

Twenty of the 32 crystal classes are piezoelectric, and crystals belonging to one of these classes (point groups) display piezoelectricity. All piezoelectric classes lack inversion symmetry. Any material develops a dielectric polarization when an electric field is applied, but a substance that has such a natural charge separation even in the absence of a field is called a polar material. Whether or not a material is polar is determined solely by its crystal structure. Only 10 of the 32 point groups are polar. All polar crystals are pyroelectric, so the 10 polar crystal classes are sometimes referred to as the pyroelectric classes.

There are a few crystal structures, notably the perovskite structure, which exhibit ferroelectric behavior. This is analogous to ferromagnetism, in that, in the absence of an electric field during production, the ferroelectric crystal does not exhibit a polarization. Upon the application of an electric field of sufficient magnitude, the crystal becomes permanently polarized. This polarization can be reversed by a sufficiently large counter-charge, in the same way that a ferromagnet can be reversed. However, although they are called ferroelectrics, the effect is due to the crystal structure (not the presence of a ferrous metal).

Corpuscular Theory of Crystals

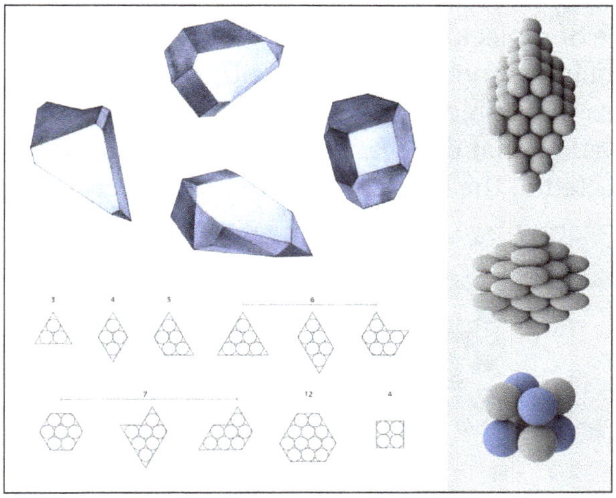

After Robert Hooke observed tiny crystals with regular-shaped faces under his microscope, he postulated that crystals were composed of identical spherical corpuscles. The ordered arrangement of these spheres gave rise to regular crystal faces. For example, three spheres can form equilateral triangles, while four can form rhombuses, and five can form isosceles trapezoids. Hooke also claimed that four spheres can form a tetrahedron, but he did not describe other possible three-dimensional structures. Wollaston further developed Hooke's theory. In a paper published in 1897, he analyzed the packing of spherical, ellipsoidal corpuscles, and corpuscles with different sizes (i.e. structures on the left).

Haüy's Theory of Crystals

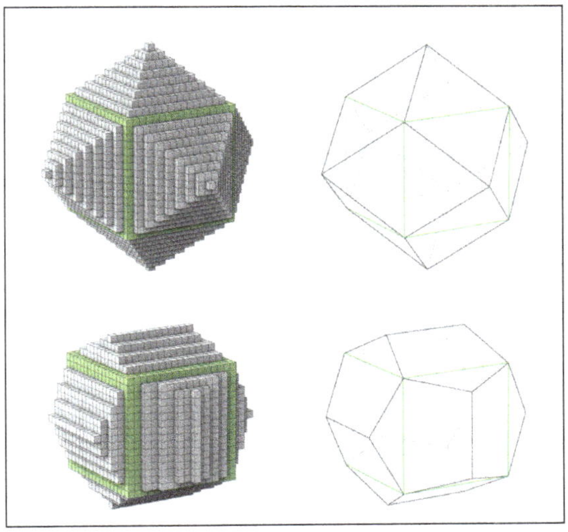

Haüy is recognized as the father of modern crystallography. He believes that crystals are made up of orderly arranged "integrant molecules" and different crystals should have different "integrant molecules". Haüy's integrant molecule is similar to the concept of unit cell in crystallography. Using his laws of decrement, he explained why the same crystal could occur in different geometries. For example, a cube-shaped crystal with cubic integrant molecules (green color in the above figure)

can be transformed to a dodecahedral crystal with rhombic faces. To do this, we can add new layers of integrant molecules on each surface of the cubic crystal (gray color in the above figure) but each time diminish one line of molecules from each side. If we just diminish one molecule for two opposite sides while diminish two molecules for the other two sides when each time we overlay a layer of integrant molecules, the cubic crystal will be transformed to a dodecahedral crystal with pentagonal faces (not an orthogonal dodecahedron). In fact, Haüy's laws were very advanced at that time and tightly connected to the lattice planes in modern crystallography.

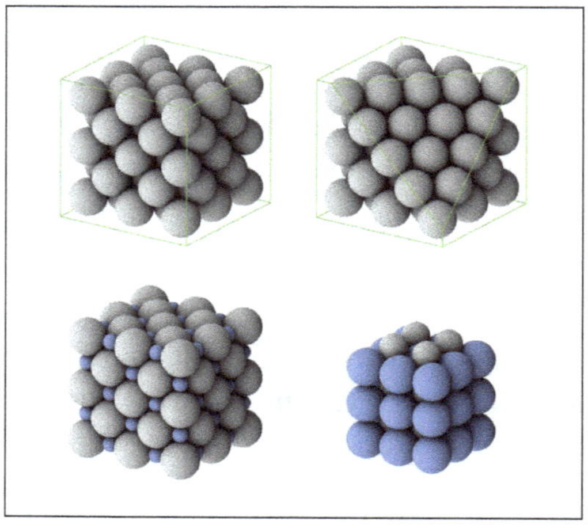

Top row: one type of close-packed structure of equal spheres. Bottom row: crystal structures similar to sodium chloride (left) and to cesium chloride (right).

Barlow's Prediction of Crystal Structures

Scientists predicted some possible crystal structures before the invention of X-ray diffraction method. One good example as a series of crystal structures published by Barlow in 1897. For a crystal that only consisted of one type of atoms, he regarded all atoms as solid spheres. As a result, the crystal structure was equivalent to the closest packing of such solid spheres. (There are actually two types of closest packing but only one of them is shown in the above figure.) Barlow further analyzed the closest packing of two types of spheres and correctly predicted the structures similar as NaCl and CsCl.

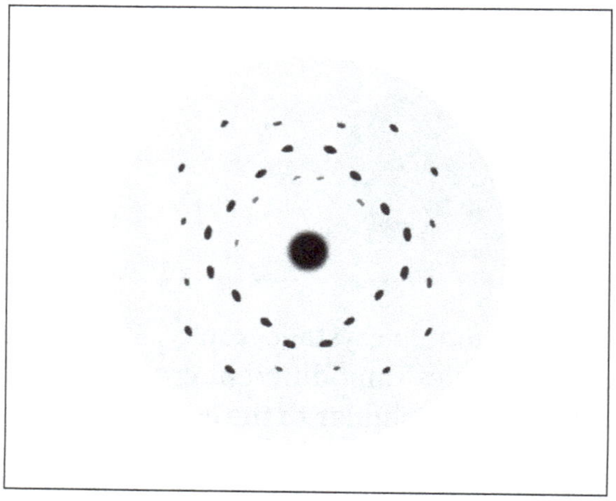

Laue's Crystal Diffraction Experiment

In 1912, 17 years after Roentgen discovered X-ray, von Laue suddenly came up with a splendid idea. He wondered if diffraction would occur when a beam of X-ray pass through a crystal since the characteristic wavelength of X-ray is close to the spacing between atoms in a crystal. In a couple of months, he proved this assumption with the help of two excellent experimentalists and they acquired the first photo of X-ray diffraction. The upper left image illustrates the X-ray diffraction photo of sphalerite took by Laue and his colleagues in 1912. Laue's experiment not only validated the electromagnetic nature of X-Ray, but provided a powerful tool for scientist to study the microscopic structures as well. Einstein highly appraised X-ray diffraction as one the greatest physics experiments. Due to his great contribution in X-Ray diffraction, Laue received Noble Prize in physics in 1914.

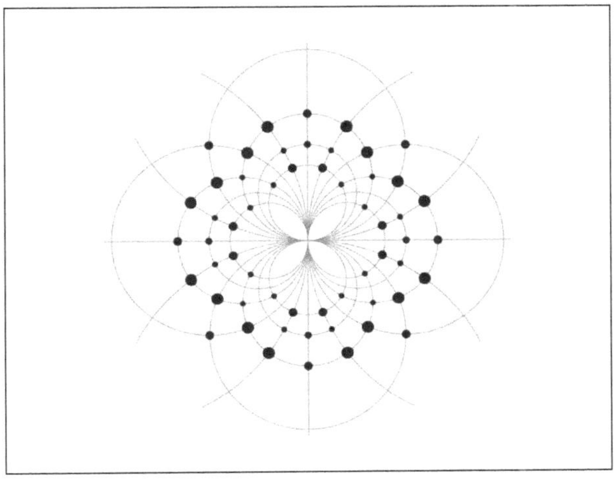

Bragg's Equation

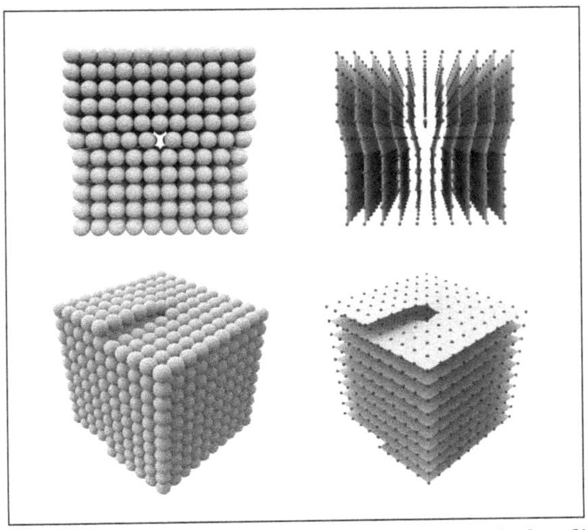

Top left: edge dislocation, top right: crystal plane schematic of edge dislocation, bottom left: screw dislocation, bottom right: crystal plane schematic of screw dislocation.

Although Laue's X-Ray diffraction experiment triggered great interests in academia, it was W. H. Bragg and his son, W. L. Bragg, who successfully explained Laue's results and made great contributions the X-ray diffraction experiments. Laue incorrectly interpreted the diffracted X-ray as an excitation consequence when the incident X-ray interacted with the crystal. However, W. H. Bragg and W. L. Bragg realized that the diffraction was actually a result of reflection of X-ray from certain crystal planes and the spacing between these planes should satisfy the famous Bragg's equation. According to this principle, they changed the configuration of the apparatus from transmission to reflection. Therefore, solving a crystal structure finally became a reality with their contributions and the structural analysis of NaCl and diamond were very significant results at that time. Afterwards, the X-ray diffraction technology was developed rapidly. Scientist now can use it to solve the structures of inorganic materials, organic molecules, DNA, and proteins (We will see the related applications later in this book). The bottom left figure is an illustration in "X Rays and Crystal Structure", which provided the theoretical positions of Laue diffraction spots for a piece of NaCl crystal.

Dislocation

There are always a variety of defects in crystal structures. We will introduce one type of defects, dislocation, which includes two basic types: edge dislocation and screw dislocation. The concept of dislocation was firstly suggested by Tyler et al. in 1934. In 1956, Hirsch et al. confirmed the existence of dislocation by the means of transmission electron microscopy. The mechanical properties of metal are affected to dislocations. When plastic deformation applied to metals, new dislocations are generated and the gliding of dislocations occurs. In addition, screw dislocations influence crystal growth and cause the helical structure that could be observed under a microscope.

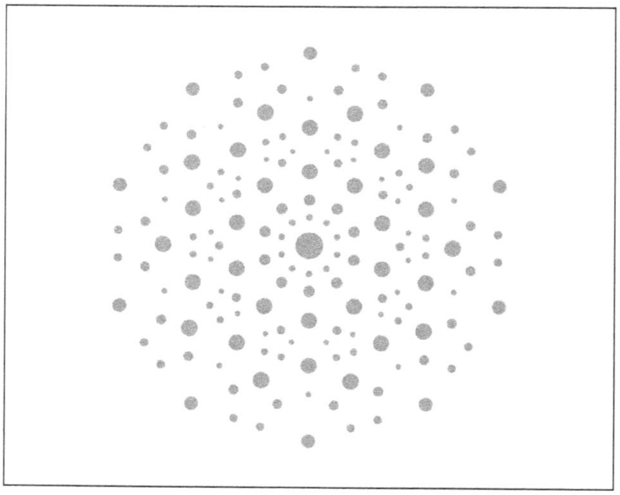

Quasicrystals

In 1982, D. Shechtman observed a diffraction pattern that he hardly believed when he was using electron diffraction to study a fast quenched Al-Mn alloy (above image). The spots were as clear as the diffraction spots from crystals, but what confused him was the decagonal symmetry, which was impossible for crystals. Shechtman immediately realized that this pattern should imply a totally new solid structure, ordered but not periodic, and which was why it still generated clear diffraction patterns. Shechtman published the result in 1984. Scientist named this type of material "quasicrystals" and realized its similarity to Penrose pattern. The discovery of quasicrystals triggered intense debates in academia. Many scientists, including L. Pauling, the two-time Nobel Laureate, denied its existence. He criticized that "No quasicrystals but only quasi-scientists." However, more and more quasicrystals were discovered. In 1992, the International Union of Crystallography revised the definition of crystals and included the quasicrystal into the new definition. In 2011, Shechtman received Nobel Prize in chemistry for his contribution in quasicrystals. The establishment of quasicrystal is a typical example of a new idea challenging old ideas in the history of science.

Crystal Systems

The crystal system is a grouping of crystal structures that are categorized according to the axial system used to describe their "lattice". A crystal's lattice is a three dimensional network of atoms that are arranged in a symmetrical pattern. Each crystal system consists of a set of three axes in a particular geometrical arrangement. The seven unique crystal systems, listed in order of decreasing symmetry, are: 1. Isometric System, 2. Hexagonal System, 3. Tetragonal System, 4. Rhombohedric (Trigonal) System, 5. Orthorhombic System, 6. Monoclinic System, 7. Triclinic System.

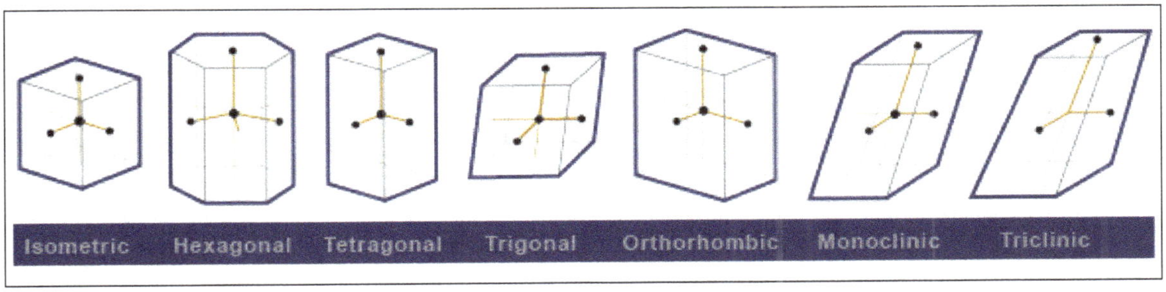

The Seven Crystal Systems

Cubic

The Cubic crystal system is also known as the "isometric" system. The Cubic (Isometric) crystal system is characterized by its total symmetry. The Cubic system has three crystallographic axes that are all perpendicular to each other and equal in length. The cubic system has one lattice point on each of the cube's four corners.

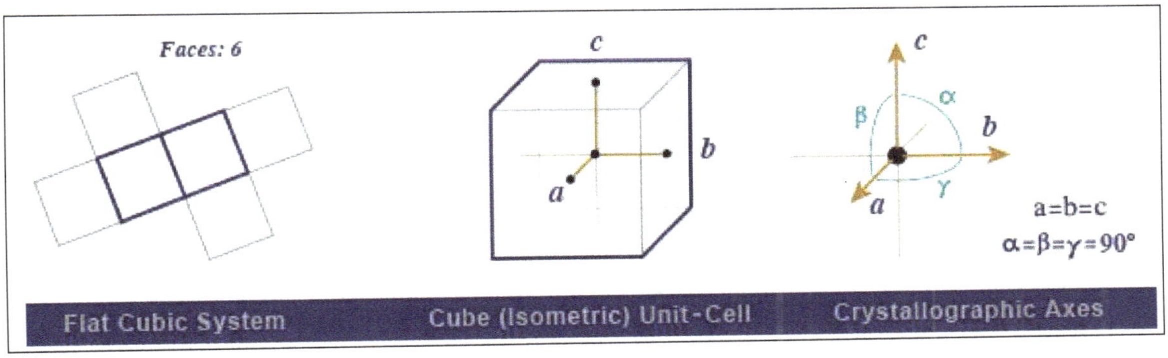

Hexagonal

The Hexagonal crystal system is has four crystallographic axes consisting of three equal horizontal or equatorial (a, b, and d) axes at 120°, and one vertical (c) axis that is perpendicular to the other three. The (c) axis can be shorter or longer than the horizontal axes.

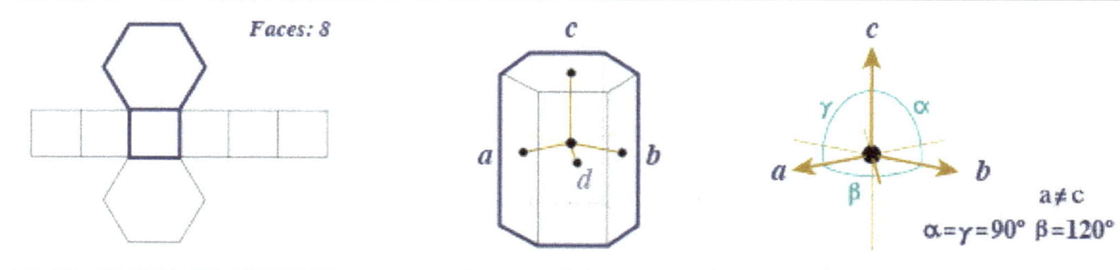

Flat Hexagonal Crystal | Hexagonal Unit-Cell | Crystallographic Axes

Tetragonal

A Tetragonal crystal is a simple cubic that is stretched along its (c) axis to form a rectangular prism. The Tetragonal crystal will have a square base and top, but a height that is taller. By continuing to stretch the "body-centered" cubic one more Bravais lattice of the Tetragonal system is constructed.

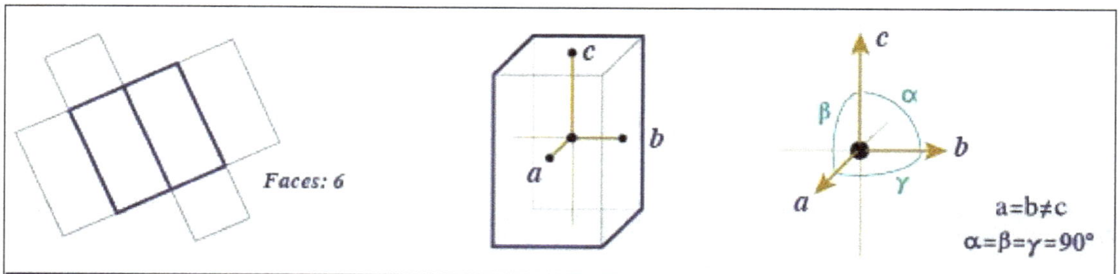

Flat Tetragonal System | Tetragonal Unit-Cell | Crystallographic Axes

Rhombohedral

A Rhombohedron (aka Trigonal) has a three-dimensional shape that is similar to a cube that has been compressed to one side. Its form is considered prismatic, as all faces are parallel to each other. The faces that are not square are called "rhombi." A rhombohedral crystal has six faces or rhombi, 12 edges, and 8 vertices. If all of the non-obtuse internal angles of the faces are equal (flat sample), it can be called a trigonal trapezohedron.

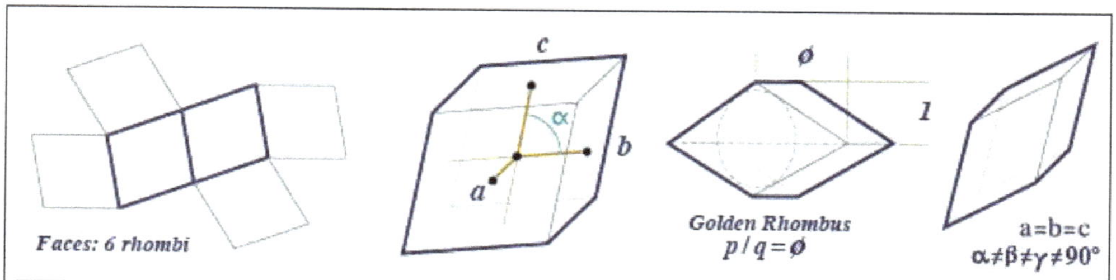

Flat Rhombohedron | Rhombohedral Unit-Cell | Crystallographic Axes

Orthorhombic

Minerals that form in the Orthorhombic (aka Rhombic) crystal system have three mutually perpendicular axes, all with different or unequal lengths.

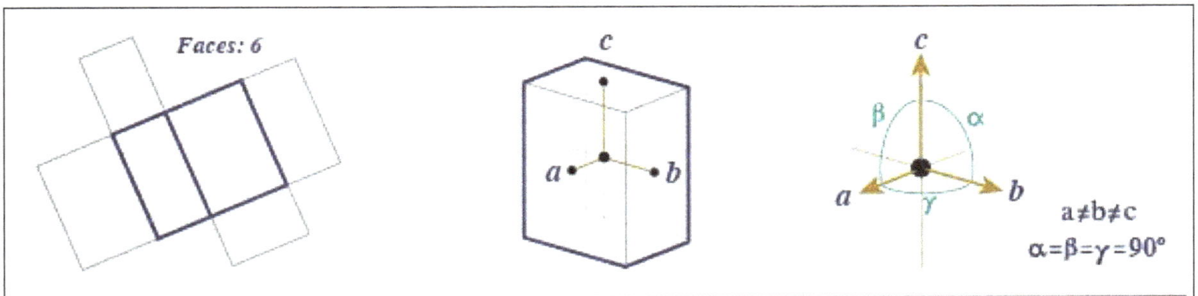

Flat Monoclinic System | Orthorhombic Unit-Cell | Crystallographic Axes

Monoclinic

Crystals that form in the Monoclinic System have three unequal axes. The (a) and (c) crystallographic axes are inclined toward each other at an oblique angle, and the (b) axis is perpendicular to a and c. The (b) crystallographic axis is called the "ortho" axis.

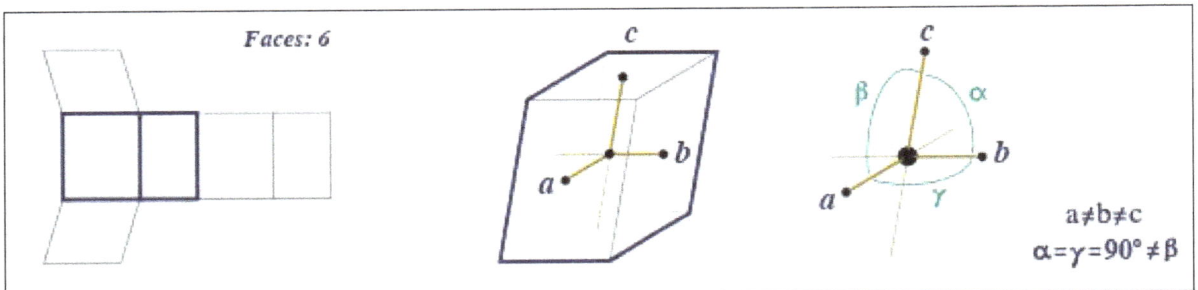

Flat Monoclinic Crystal | Monoclinic Unit-Cell | Crystallographic Axes

Triclinic

Crystals that form in the Triclinic System have three unequal crystallographic axes, all of which intersect at oblique angles. Triclinic crystals have a 1-fold symmetry axis with virtually no symmetry and no mirrored planes.

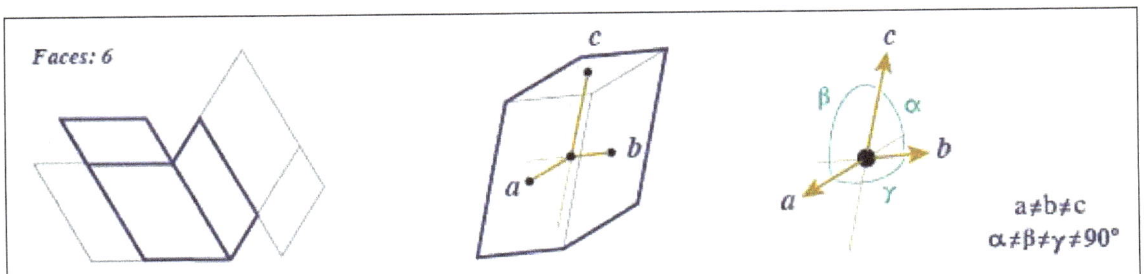

Flat Triclinic Crystal | Triclinic Unit-Cell | Crystallographic Axes

Crystal Forms

A crystal form is a set of faces which are geometrically equivalent and whose spatial positions are related to one another according to the symmetry of the crystal. If one face of a crystal form is defined, the point symmetry operations which specify the class to which the crystal belongs also determine the other faces of the crystal form.

Fifteen different forms are possible within the isometric or cubic system. These include the hexoctahedron, gyroid, hextetrahedron, diploid, and tetartoid, among others. The crystal forms of the remaining five crystal systems are the monohedron or pedion, parallelohedron or pinacoid, dihedron, or dome and sphenoid, disphenoid, prism, pyramid, dipyramid, trapezohedron, scalenohedron, rhombohedron, and tetrahedron.

The crystal forms which occur in each crystal class and system must possess a symmetry complementary to that of the associated crystal class and system. For example, a monohedron, which possesses only one face, will never occur in a crystal with inversion symmetry because the inversion operation requires that an equivalent face be present on the opposite side of the crystal.

A simple crystal may consist of only a single crystal form. A more complicated crystal may be a combination of several different forms. All forms which occur in a crystal of a particular system must be compatible with that crystal system.

Monohedron

The monohedral crystal form is also called a pedion. It consists of a single face which is geometrically unique for the crystal and is not repeated by any set of symmetry operations. Members of the triclinic crystal system produce monohedral crystal forms.

Parallelohedron

The parallelohedral crystal form is also called a pinacoid. It consists of two and only two geometrically equivalent faces which occupy opposite sides of a crystal. The two faces are parallel and are related to one another only by a reflection or an inversion. Members of the triclinic crystal system produce parallelohedral crystal forms.

Dihedron

The dihedron consists of two and only two nonparallel geometrically equivalent faces. The two faces may be related by a reflection or by a rotation. The dihedron is termed a dome if the two faces are related only by reflection across a mirror plane. If the two faces are related instead by a 2-fold rotation axis then the dihedron is termed a sphenoid. Members of the monoclinic crystal system produce dihedral crystal forms.

Disphenoid

Members of the orthorhombic and tetragonal crystal systems produce rhombic and tetragonal disphenoids, which possess two sets of nonparallel geometrically equivalent faces, each of which is related by a 2-fold rotation. The faces of the upper sphenoid alternate with the faces of the lower sphenoid in such forms.

Prism

A prism is composed of a set of 3, 4, 6, 8, or 12 geometrically equivalent faces which are all parallel to the same axis. Each of these faces intersects with the two faces adjacent to it to produce a set of parallel edges. The mutually parallel edges of all intersections of the prism sides then form a tube. Prisms are given names based on the shape of their cross section. Variants of the prism form include the rhombic prism, tetragonal prism, trigonal prism, and hexagonal prism. A prism in which the large faces are divided into two mirror-image faces which intersect with one another at an oblique angle is called a ditetragonal prism, a ditrigonal prism, or a dihexagonal prism. Prisms are associated with the members of the monoclinic crystal system.

Pyramid

A pyramid is composed of a set of 3, 4, 6, 8, or 12 faces which are not parallel but instead intersect at a point. The orthorhombic, tetragonal and hexagonal crystal systems all produce pyramids. These pyramids are named according to the shape of their cross-section in the same way that prisms are. Thus are produced the rhombic pyramid, tetragonal pyramid, trigonal pyramid, and hexagonal pyramid. Each large face of the ditetragonal pyramid, ditrigonal pyramid, and dihexagonal pyramids is divided into two mirror-image faces which occupy an oblique angle with respect to one another.

Dipyramid

The dipyramidal crystal form is composed of two pyramids placed base-to-base and related by reflection across a mirror plane which runs parallel to and adjacent to the pyramid bases. The upper and lower pyramids may each have 3, 4, 6, 8, or 12 faces; the dipyramidal form therefore possesses a total of 6, 8, 12, 16, or 24 faces. The orthorhombic, tetragonal and hexagonal crystal systems all produce dipyramids. These dipyramids are named for the shape of their cross-section just as prisms and pyramids are, resulting in the rhombic dipyramid, trigonal dipyramid, tetragonal dipyramid, and hexagonal dipyramid. The large faces of the ditetragonal, ditrigonal and dihexagonal dipyramids are divided into two mirror-image faces which intersect one another at an oblique angle.

Trapezohedron

A trapezohedron is a crystal form possessing 6, 8, or 12 trapezoidal faces. The tetragonal crystal system and both the trigonal and hexagonal divisions of the hexagonal crystal system produce trapezohedral crystal forms. Trigonal trapezohedra possess three trapezoidal faces on the top and three on the bottom for a total of six faces; tetragonal trapezohedra have four faces on top and four on the bottom for a total of eight faces; and hexagonal trapezohedra have six faces on top and six on the bottom, resulting in twelve faces total.

Scalenohedron

A scalenohedron consists of 8 or 12 faces, each of which is a scalene triangle. The faces appear to be grouped into symmetric pairs. The tetragonal and hexagonal crystal systems produce the scalenohedral crystal form, of which examples may be further described as trigonal, tetragonal and hexagonal scalenohedra.

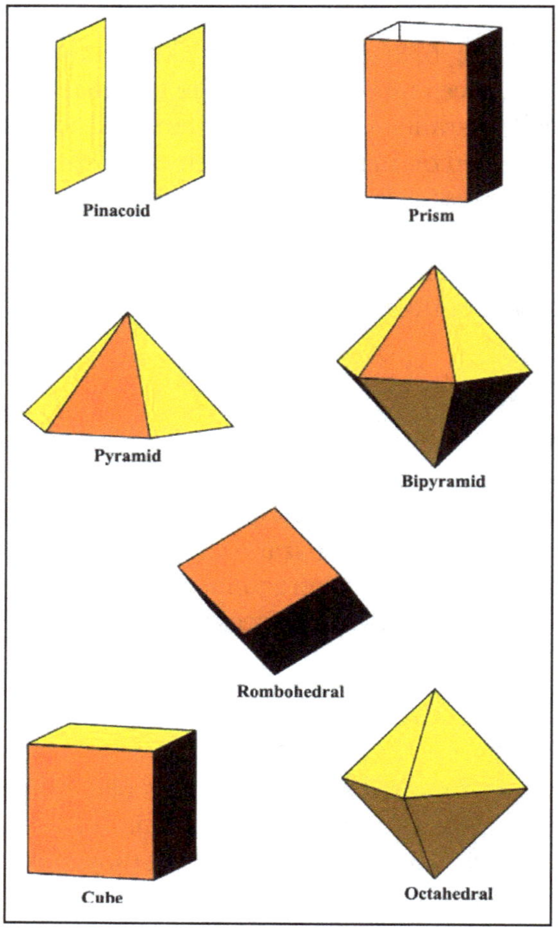

Rhombohedron

The rhombohedral crystal form possesses six rhombus-shaped faces. A rhombohedron resembles in appearance a cube which is poised upright upon one corner and has been either flattened or elongated along an axis which runs diagonally from corner to corner through the center. The rhombohedral crystal form is produced only by members of the trigonal and rhombohedral divisions of the hexagonal crystal system.

Tetrahedron

A tetrahedron is composed of four triangular faces. In crystals of the isometric system each face is an identical equilateral triangle. In crystals of the tetragonal system each face is an identical isoceles triangle; this variant of the tetrahedron is called a tetragonal tetrahedron. In crystals of the orthorhombic system the faces consist of two pairs of different isoceles triangles; the crystal is then termed a rhombic tetrahedron.

References

- Crystal-structure: beautifulchemistry.net, Retrieved 23 March, 2019
- Hook, J.R.; Hall, H.E. (2010). Solid State Physics. Manchester Physics Series (2nd ed.). John Wiley & Sons. ISBN 9780471928041
- Schwartz, Mel (2002). "Tin and Alloys, Properties". Encyclopedia of Materials, Parts and Finishes (2nd ed.). CRC Press. ISBN 978-1-56676-661-6
- Kallay, Nikola (2000). Interfacial Dynamics. CRC Press. ISBN 978-0824700065

5

Crystal Defects

In crystallography, crystal defects are the discrepancies in the regular pattern, arrangement and organization of atoms of crystalline solids. Some of the defects are intrinsic and extrinsic defects, Frenkel defects, interstitial defects, line defects, surface defects, volume defects, etc. The topics elaborated in this chapter will help in gaining a better perspective about these crystal defects.

A crystal is never perfect; a variety of imperfections can mar the ordering. A defect is a small imperfection affecting a few atoms. The simplest type of defect is a missing atom and is called a vacancy. Since all atoms occupy space, extra atoms cannot be located at the lattice sites of other atoms, but they can be found between them; such atoms are called interstitials. Thermal vibrations may cause an atom to leave its original crystal site and move into a nearby interstitial site, creating a vacancy-interstitial pair. Vacancies and interstitials are the types of defects found in a pure crystal. In another defect, called an impurity, an atom is present that is different from the host crystal atoms. Impurities may either occupy interstitial spaces or substitute for a host atom in its lattice site.

There is no sharp distinction between an alloy and a crystal with many impurities. An alloy results when a sufficient number of impurities are added that are soluble in the host metal. However, most elements are not soluble in most crystals. Crystals generally can tolerate a few impurities per million host atoms. If too many impurities of the insoluble variety are added, they coalesce to form their own small crystallite. These inclusions are called precipitates and constitute a large defect.

Germanium is a common impurity in silicon. It prefers the same tetrahedral bonding as silicon and readily substitutes for silicon atoms. Similarly, silicon is a common impurity in germanium. No large crystal can be made without impurities; the purest large crystal ever grown was made of germanium. It had about 10^{10} impurities in each cubic centimetre of material, which is less than one impurity for each trillion atoms.

Impurities often make crystals more useful. In the absence of impurities, α-alumina is colourless. Iron and titanium impurities impart to it a blue colour, and the resulting gem-quality mineral is known as sapphire. Chromium impurities are responsible for the red colour characteristic of rubies, the other gem of α-alumina. Pure semiconductors rarely conduct electricity well at room temperatures. Their ability to conduct electricity is caused by impurities. Such impurities are deliberately added to silicon in the manufacture of integrated circuits. In fluorescent lamps the visible light is emitted by impurities in the phosphors (luminescent materials).

Other imperfections in crystals involve many atoms. Twinning is a special type of grain boundary defect, in which a crystal is joined to its mirror image. Another kind of imperfection is a dislocation, which is a line defect that may run the length of the crystal. One of the many types of dislocations is due to an extra plane of atoms that is inserted somewhere in the crystal structure. Another type, called an edge dislocation, is shown in figure. This line defect occurs when there is a missing row of atoms. In the figure the crystal arrangement is perfect on the top and on the bottom. The defect is the row of atoms missing from region b. This mistake runs in a line that is perpendicular to the page and places a strain on region *a*.

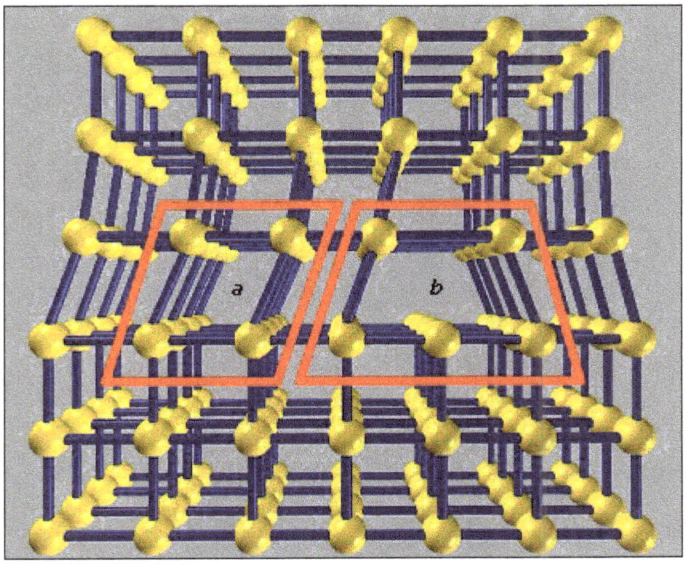

Crystalline lattice defect. An edge dislocation occurs when there is a missing row of atoms as shown in region b. Region *a* is strained.

Dislocations are formed when a crystal is grown, and great care must be taken to produce a crystal free of them. Dislocations are stable and will exist for years. They relieve mechanical stress. If one presses on a crystal, it will accommodate the induced stress by growing dislocations at the surface, which gradually move inward. Dislocations make a crystal mechanically harder. When a metal bar is cold-worked by rolling or hammering, dislocations and grain boundaries are introduced; this causes the hardening.

Point Defects

Point defects are where an atom is missing or is in an irregular place in the lattice structure. Point defects include self-interstitial atoms, interstitial impurity atoms, substitutional atoms and vacancies. A self-interstitial atom is an extra atom that has crowded its way into an interstitial void in the crystal structure. Self-interstitial atoms occur only in low concentrations in metals because they distort and highly stress the tightly packed lattice structure.

A substitutional impurity atom is an atom of a different type than the bulk atoms, which has replaced one of the bulk atoms in the lattice. Substitutional impurity atoms are usually close in size (within approximately 15%) to the bulk atom. An example of substitutional impurity atoms is the zinc atoms in brass. In brass, zinc atoms with a radius of 0.133 nm have replaced some of the copper atoms, which have a radius of 0.128 nm.

Interstitial impurity atoms are much smaller than the atoms in the bulk matrix. Interstitial impurity atoms fit into the open space between the bulk atoms of the lattice structure. An example of interstitial impurity atoms is the carbon atoms that are added to iron to make steel. Carbon atoms, with a radius of 0.071 nm, fit nicely in the open spaces between the larger (0.124 nm) iron atoms.

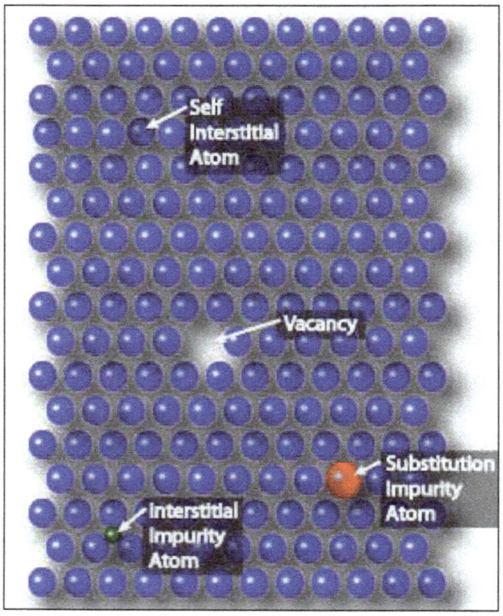

Vacancies are empty spaces where an atom should be, but is missing. They are common, especially at high temperatures when atoms are frequently and randomly change their positions leaving behind empty lattice sites. In most cases diffusion (mass transport by atomic motion) can only occur because of vacancies.

Intrinsic Defects

An intrinsic defect is formed when an atom is missing from a position that ought to be filled in the crystal, creating a vacancy, or when an atom occupies an interstitial site where no atom would ordinarily appear, causing an interstitialcy.

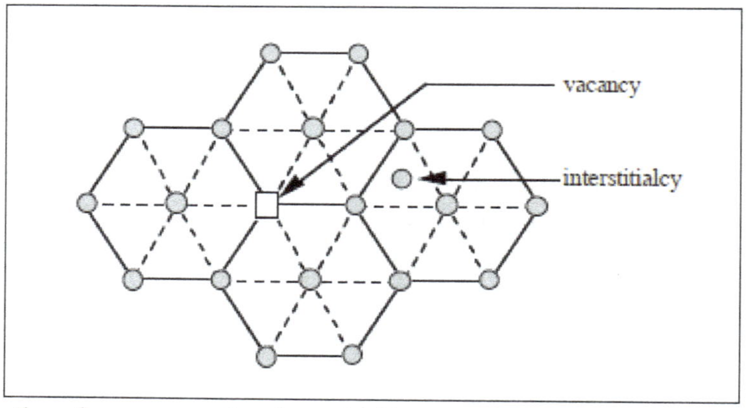

Illustration of a vacancy and an interstitial in a two-dimensional hexagonal lattice.

Because the interstitial sites in most crystalline solids are small (or have an unfavorable bonding configuration, as, for example, in the diamond lattice) interstitialcies are high-energy defects that are relatively uncommon. Vacancies, on the other hand, are present in a significant concentration in all crystalline materials. Their most pronouncedeffect is to govern the migration of atoms on the crystal lattice (solid state diffusion). In order for an atom to move easily from one crystal lattice site to another the target site must be vacant. As we shall see, the rate of diffusion on the crystal lattice is largely governed by the concentration of vacancies.

Ordered compounds can have more complex intrinsic defects. In most compounds the different species are charged to at least some degree. An intrinsic defect destroys the local charge balance, which must be restored in some way. The compound defects that preserve charge are easiest to visualize in binary ionic solids like NaCl. An isolated vacancy in an ionic solid creates an excess charge. The excess charge can be compensated by a paired vacancy on the sublattice of the other specie; for example, the excess charge associated with a Na vacancy is balanced if there is a Cl vacancy nearby.

Schottky Defects

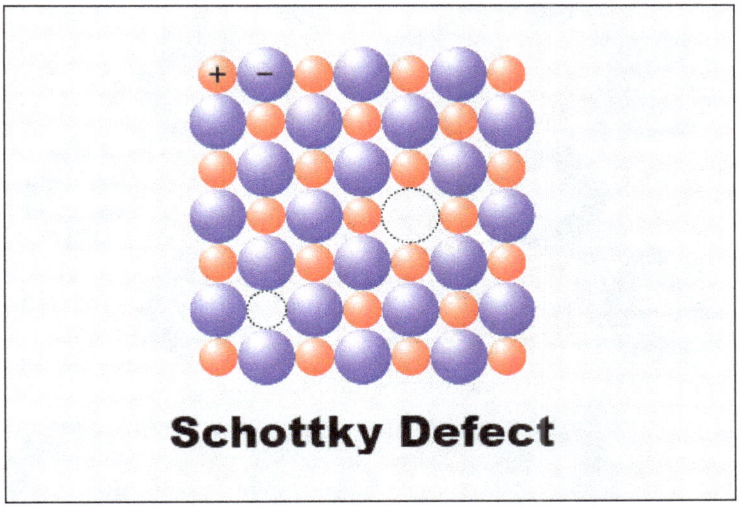

Schottky defect is a type of point defect or imperfection in solids which is caused by a vacant position that is generated in a crystal lattice due to the atoms or ions moving out from the interior to the surface of the crystal.

Characteristics of Schottky Defect

Some of the distinct characteristics of this defect are:

- There is a very small difference in size between cation and anion.
- Cation and anion both leave the solid crystal.
- Atoms also move out of the crystal permanently.
- Generally two vacancies are formed.
- As for the density of the solid it decreases considerably.

In ionic crystals, this point defect forms when oppositely charged ions leave their lattice sites, creating vacancies. These vacancies are formed in stoichiometric units, to maintain an overall neutral charge in the ionic solid. The surrounding atoms then move to fill these vacancies, causing new vacancies to form. Normally these defects will lead to a decrease in the density of the crystal or metal. Chemical equations in Kröger–Vink notation for the formation of Schottky defects in TiO_2 and $BaTiO_3$.

$$\varnothing \rightleftarrows v_{Ti}^{''''} + 2v_{O}^{\bullet\bullet}$$
$$\varnothing \rightleftarrows v_{Ba}^{''} + v_{Ti}^{''''} + 3v_{O}^{\bullet\bullet}$$

This can be illustrated schematically with a two-dimensional diagram of a sodium chloride crystal lattice:

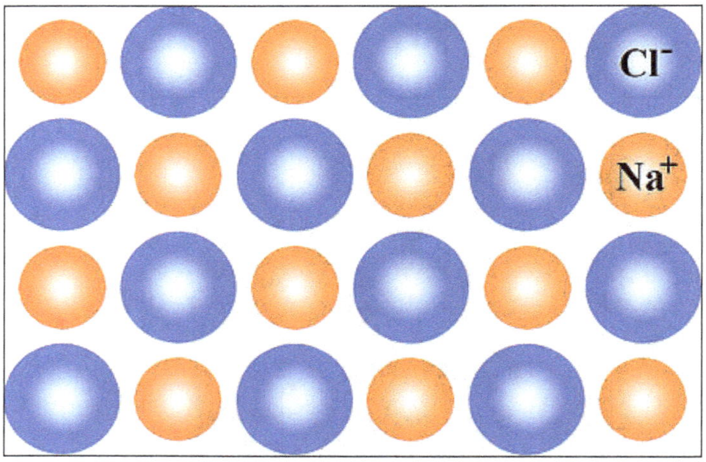

The defect-free NaCl structure.

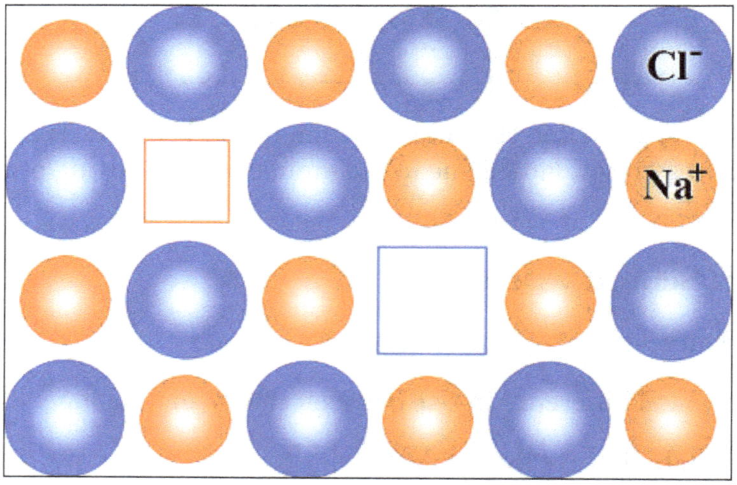

Schottky defects within the NaCl structure.

Schottky defects consist of unoccupied anion and cation sites in a stoichiometric ratio. For a simple ionic crystal of type A^-B^+, a Schottky defect consists of a single anion vacancy (A) and a single cation vacancy (B), or $v_A^{\bullet} + v_B^{'}$ following Kröger–Vink notation. For a more general crystal with formula A_xB_y, a Schottky cluster is formed of x vacancies of A and y vacancies of B, thus the overall stoichiometry and charge neutrality are conserved.

Schottky defects are observed most frequently when there is a small difference in size between cations and anions. This is produced as the result of the thermal incorporation of unoccupied lattice sites from the exterior of the crystal. The lattice undergoes thermal vibration and thermal expansion when the temperature is raised above 0 K. When it happens the pair of vacancies is incorporated in the crystal. So electrical neutrality is maintained inside the crystals.

Examples:

It is a type of defect in crystals that mostly occurs in highly ionic compounds or highly coordinated compounds. The compound lattice has only a small difference in sizes between the anions and cations. Some of the common example of salts where Schottky defect is prominent includes Sodium Chloride (NaCl), Potassium Chloride (KCl), Potassium Bromide (KBr), Caesium Chloride (CsCl) and Silver Bromide (AgBr).

Calculation:

The number of defects for an MX crystal can be calculated by the given formula:

$$n_s \approx N \exp\left(-\frac{\Delta H_s}{2RT}\right)$$

Where,

- n_s = number of Schottky defects per unit volume,
- ΔH = enthalpy of defect formation,
- R = gas constant,
- T = absolute temperature (in K).

We can calculate N by using the formula;

$$N = \frac{\text{Density of the ionic crystal compound} \times \text{NA}}{\text{Molar mass of the ionic crystal compound}}$$

Frenkel Defects

Frenkel Defect is one type of Point Defect; in fact, it is a combination of both Vacancy and Interstitial type of point defects. Usually, this type of defect is observed in ionic solids, where size of anion is substantially larger than the size of cation. Basically, a Frenkel Defect is one type of point defect where an atom (better to say ion, especially cation) leaves its original lattice site and occupies an interstitial position on the same crystal.

When an atom leaves the original lattice site then Vacancy Defect is created. Alternatively, when an atom (of same crystal or foreign crystal) occupies the interstitial site then Interstitial Defect is created. Since, in Frenkel Defect, one atom leaves the original lattice site and occupies an interstitial site in the same crystal, so Frenkel Defect is a pair of Vacancy & Self-interstitial Defects. The following image depicts a typical crystal with Frenkel Defect.

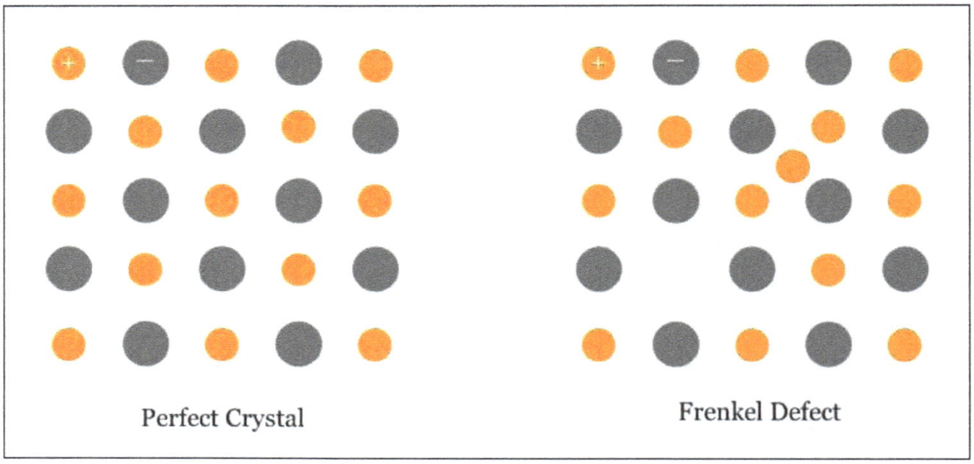

In Frenkel Defect, cations move from original lattice site to interstitial position; while, the larger ions (anion) remain in their corresponding lattice sites.

Example of Materials Where Frenkel Defect can be found:

- Zinc Sulfide (ZnS),
- Silver Chloride (AgCl),
- Silver Bromide (AgBr).

Features of Frenkel Defect

- In Frenkel Defect, neither any atom leaves the solid crystal nor do any foreign atoms occupy any interstitial position. So the overall density of the crystal solid remains unchanged.
- This type of defect is usually observed in ionic crystals, where size of anion is substantially larger than the size of cation. For example: zinc sulfide (ZnS), silver chloride (AgCl), etc.
- The smaller ions, usually cation, move from original lattice site to interstitial position; while, the larger ions (anion) remain in their corresponding lattice sites.
- It does not affect the chemical property of the solid.
- Electrical neutrality is also maintained in this defect.

Calculation:

Frenkel Defects can be calculated using the equation given below:

$$\sqrt{NN^*}\, e^{\delta \frac{H}{2RT}}$$

Where,

- N = normally occupied positions.
- N^* = number of available positions.

- Delta H = the enthalpy formation of one Frenkel defect.
- R = gas constant.

Table: Differences between Schottky defect and Frenkel defect.

Schottky Defect	Frenkel Defect
Schottky defect occurs in those ionic crystals where difference in size between cation and anion is small.	Frenkel defect usually occurs in those ionic crystals where size of anion is quite large as compared to that of the cation.
In Schottky defect, both cation and anion leave the solid crystal.	In Frenkel defect, only the smaller ion (cation) leaves its original lattice site; whereas, the anion remains in original lattice sites.
The atoms permanently leave the crystal.	Here, atoms leave the original lattice site and occupy interstitial position. So atoms reside within the solid crystal.
One Schottky defect leads to the formation of two vacancies.	One Frenkel defect creates one vacancy and one self-interstitial defect.
Two atoms reduce from the crystal for each Schottky defect.	The number of atoms present in the crystal before and after Frenkel defect remains same.
Due to vacancy formation, Schottky defect reduces density of the solid.	Density of the solid crystal before and after Frenkel defect remains same as no atom leaves the solid.
Common materials where Schottky defect can be found are: • Sodium Chloride (NaCl) • Potassium Chloride (KCl) • Potassium Bromide (KBr) • Silver Bromide (AgBr) • Cerium Dioxide (CeO_2) • Thorium Dioxide (ThO_2)	Common materials where Frenkel defect can be found are: • Zinc Sulfide (ZnS) • Silver Chloride (AgCl) • Silver Bromide (AgBr)

Interstitial Defects

An interstitial defect occurs when an atom takes the interstitial position of the lattice structure. This interstitial atom may be of the same crystal or of a foreign material. Accordingly, interstitial defect can be of two types:

- Self-Interstitial Defect: Occurs when atom of the same crystalline solid occupies the interstitial position leaving its original lattice site.
- Interstitial Defect: Occurs when a foreign atom occupies the interstitial position.

Although extra atom occupies the empty interstitial space, the size of the atom is usually larger than that of the empty space. Thus the surrounding atoms are compressed and distorted. Presence of substantial number of interstitial atoms can change the mechanical and thermal properties of the solid. However, this is sometime beneficial, and thus interstitial defects can be applied in a controlled way to enhance various properties of the solid. For example, in steel production, carbon is added with iron.

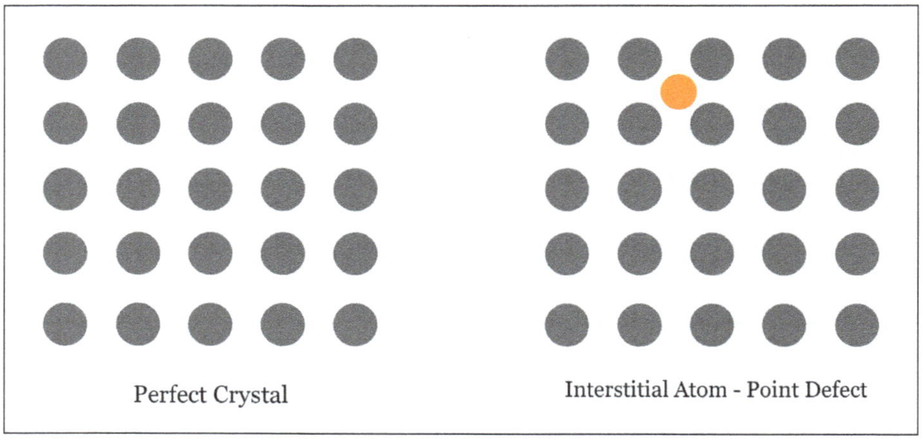

Interstitial Atom is a Point Defect. Here a foreign atom occupies the interstitial position of a perfect crystal.

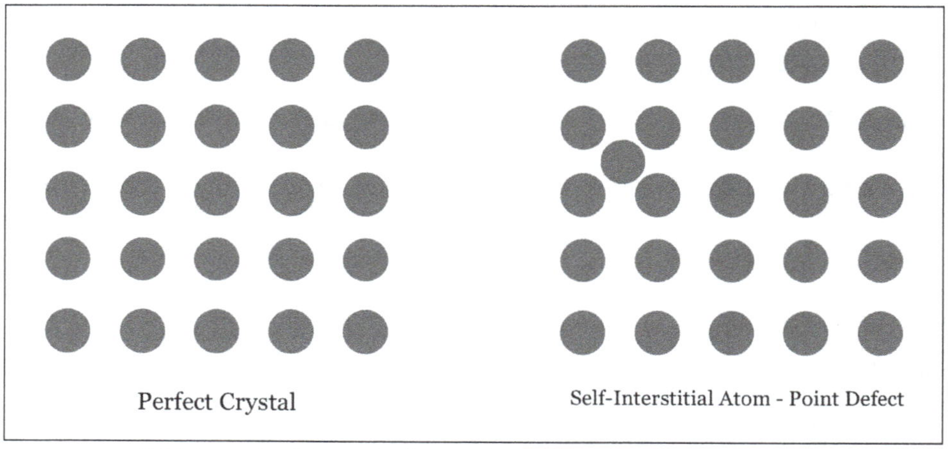

Self-Interstitial Atom is also a Point Defect. Here an extra atom of same crystal occupies the interstitial position of the crystal.

- During iron to steel production, carbon atoms occupy interstitial positions of the iron crystal.
- Similar thing happens in many alloys.
- In semiconductor materials.

An interstitial defect is formed when a foreign (solute) atom is positioned in the crystal structure at a point that is normally unoccupied. The defect is formed when a solute atom such as an alloying or impurity element sits within a gap between the crystal lattice points of the base metal (solvent). An interstitial atom is usually smaller than the solvent atoms located at the lattice points, but is larger than the interstitial site it occupies. Consequently, the surrounding crystal structure is distorted. An interstitial defect is often formed in metal alloys when the size of the solute atom is less than about 85% of the size of the host metal atom. In many cases, the solute atoms are less than half as small as the base metal. Carbon in iron (steel) is one example of an element that is interstitial.

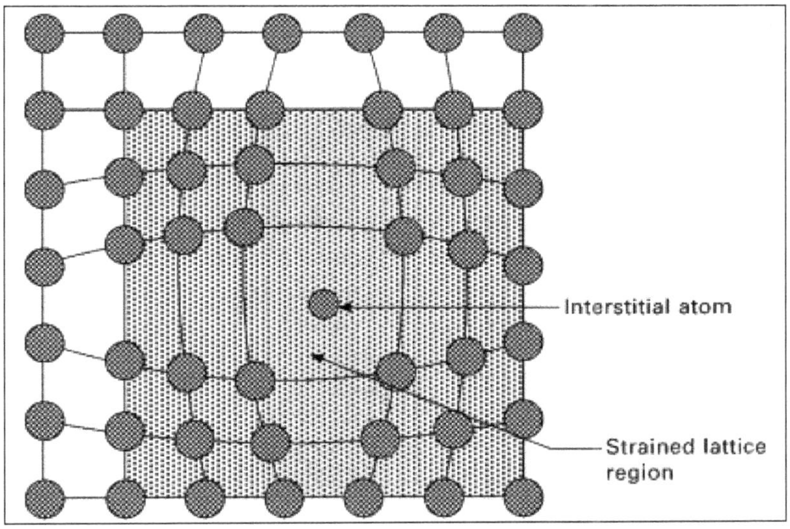

Interstitial defect within a crystal structure. The shaded region indicates the region of high strain caused by the interstitial defect.

Causes of Interstitial Defect in Solids

- Presence of foreign atoms as natural impurities within the solid.
- Deliberate addition, such as during iron to steel conversion, heat treatment, coating, etc.
- Diffusion, caused by close contact between two different materials.

Effects of Interstitial Defects

- Although extra atom occupies the empty interstitial space, the size of the atom is usually larger than that of the empty space. Thus the surrounding atoms are compressed and distorted.
- Presence of substantial number of interstitial atoms can change the mechanical and thermal properties of the solid. However, this is sometime beneficial, and thus interstitial defects can be applied in a controlled way to enhance various properties of the solid.
- Presence of large number of interstitial atoms can even convert electrically non-conductive material to conductive one.

Extrinsic Defects

The extrinsic point defects are foreign atoms, which are called solutes if they are intentionally added to the material and are called impurities if they are not. The foreign atom may occupy lattice sites, in which case it is called a substitutional solute (or impurity) or it may fill an interstitial site, in which case it is called an interstitial solute. Since the interstitial sites are relatively small, the type of the solute is largely determined by its size. Small atoms, such as hydrogen, carbon and nitrogen are often found in interstitial sites. Larger atoms are usually substitutional.

More complex extrinsic defects appear in compounds. If the valence of a substitutional defect in an ionic solid differs from that of the lattice ion then the excess charge is often compensated by a paired vacancy or interstitial. For example, when Mg^{++} ions are substituted for Na^{+} in NaCl they tend to be paired with vacancies on the Na sublattice to maintain local charge neutrality. In semiconductors substitutional atoms with the wrong valence acts as electron donors or acceptors.

Extrinsic point defects affect almost all engineering properties, but they are particularly important in semiconducting crystals, where extrinsic defects are used to control electrical properties, and in structural metals and alloys, where extrinsic defects are added to increase mechanical strength. While these properties will be discussed later in the course, it is perhaps useful to identify the characteristics of the point defects that affect them.

Line Defects

Dislocations are linear defects; they are lines through the crystal along which crystallographic registry is lost. Their principle role in the microstructure is to control the yield strength and subsequent plastic deformation of crystalline solids at ordinary temperatures. Dislocations also participate in the growth of crystals and in the structures of interfaces between crystals. They act as electrical defects in optical materials and semiconductors, in which they are almost always undesirable.

The concept of a dislocation in a solid was introduced by Volterra in the nineteenth century. However, it was not until much later that their relevance to the deformation of crystals was recognized; the notion of a dislocation as the carrier of plastic deformation did not appear until 1934. Since the 1950's it has been possible to observe and study dislocations directly using such techniques as transmission electron microscopy and x-ray topography. While dislocations influence many aspects of physical behavior, they are studied almost exclusively in Materials Science. Most of you are encountering the concept of a dislocation for the first time.

The Edge Dislocation

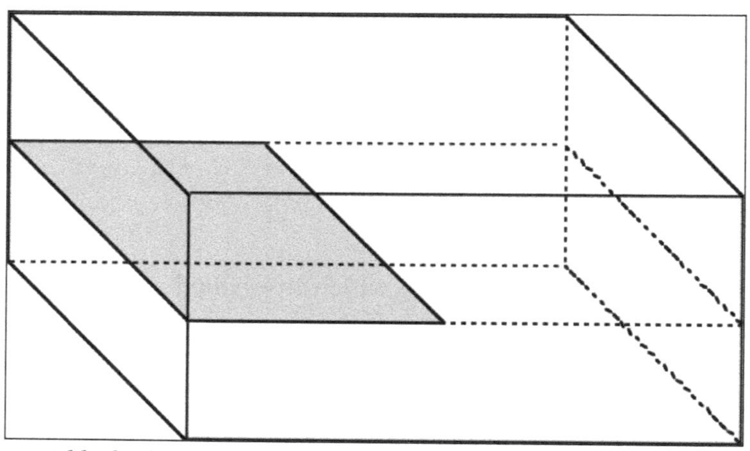

A block of material with a planar cut indicated by the shading.

The simplest way to grasp the idea of a dislocation is to imagine how you might go about creating one. We begin by making an edge dislocation, which is the easiest type of dislocation to visualize in a crystal, and follow the recipe laid down by Volterra.

Consider the solid body that is drawn in figure. For the present purpose it does not matter whether the body is crystalline; it may be easier to imagine that it is rubber.

To create an edge dislocation in this body we first make a planar cut part way through it, as illustrated by the shaded region in the figure. We then fix the part of the body below the cut, and apply a force to the body above the cut that tends to displace it in the direction of the cut, as illustrated in figure. The upper part slides, or slips over the lower by the vector distance b, which is the relative displacement of the two lips of the cut. The plane of the cut, where slip occurs, is called the slip plane. The cut is finite and constrained at its end, so material accumulates there. The end of the cut, or equivalently, the boundary of the planar region of slip, is a linear discontinuity in the material. The situation after the slip is shown in figure.

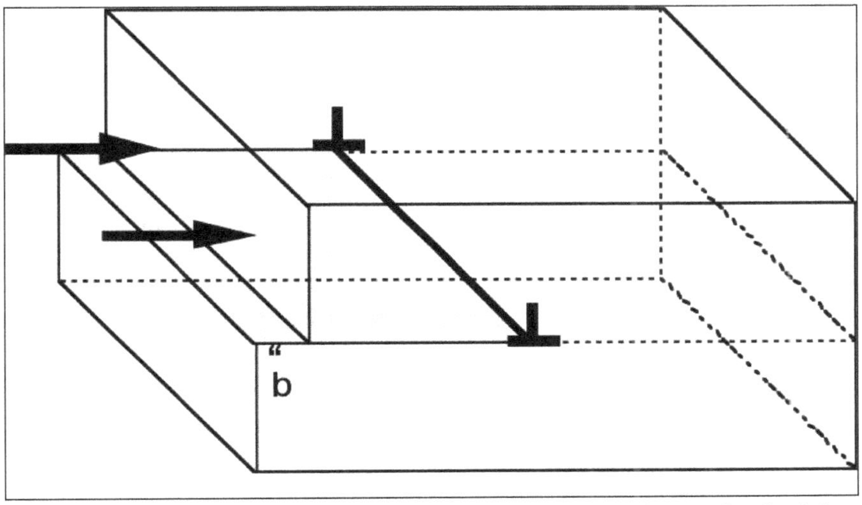

The upper part of the body has been slipped by the vector b over the shaded area. The terminal line is a discontinuity, marked by the heavy line.

Now suppose that we have a mechanism for welding the material back together which is so efficient that it is impossible to tell after the fact that the weld was ever made. If we match the material above the shaded plane to that below so that there is no physical discontinuity across the plane and weld the lips together, the shading disappears since it is impossible to tell that the material was ever separated. However, matching the material across the plane of slip requires that excess material be gathered at the line at which slip terminates. This line is, therefore, a linear defect in the material. It is called an edge dislocation. It is an isolated defect, as shown in figure since after the material has been re-welded there is no unique way to determine how the dislocation was created. For example, the dislocation would be exactly the same if the material to the right of it on the slip plane were slipped by the vector -b on the same plane. Only the exterior step indicates the origin of the dislocation, and this may be removed, or may have pre-existed the formation of the dislocation.

However it was created, the edge dislocation in figure has the property that it defines an element of slip, b, where the vector b is called the Burgers vector of the dislocation. We can always identify the slip plane of a dislocation like that shown in figure. It is the plane that contains the Burgers vector, b, and the line of the dislocation. However the dislocation actually came to its present position, its net effect is that the material above the slip plane to the left of the dislocation (in the direction of -b) has been displaced by b relative to that below the slip plane. The dislocation is a linear defect whose location is defined by its line and whose nature is characterized by its Burgers vector, b. In the case shown in figure the Burgers vector is perpendicular to the dislocation line. This perpendicularity is characteristic of an edge dislocation.

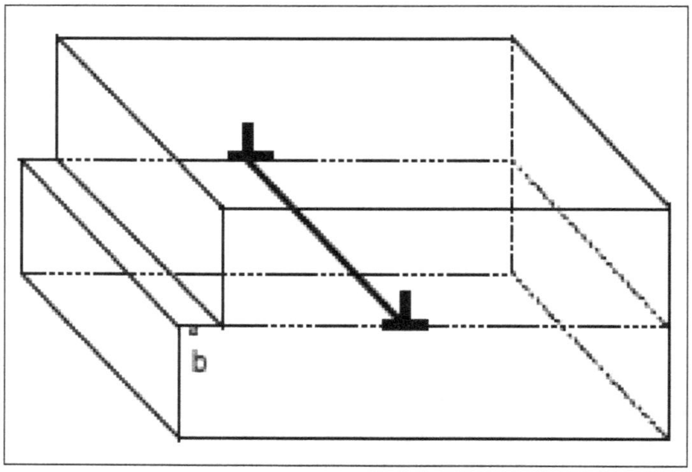

Isolated edge dislocation after the cut surface has been rejoined.

If a dislocation moves the area that has been slipped grows or shrinks accordingly. Imagine that the dislocation is initially created at the left edge of the slip plane in figure, and is then gradually moved to the right edge. Applying the construction in figure to the initial and final positions of the dislocation, it follows that the motion of the dislocation through the body causes the whole volume of material above the slip plane to be displaced by the vector b with respect to that below it, as shown in figure.

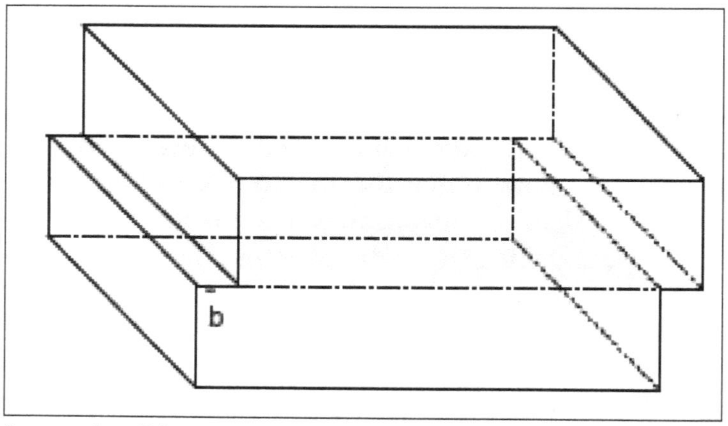

Final state of the body after an edge dislocation with Burgers' vector b has crossed the whole of the slip plane shown.

Figure illustrates the connection between the motion of a dislocation (in this case, an edge dislocation) and the plastic, or permanent deformation of a material. As we shall discuss in more detail when we consider the mechanical properties of materials, plastic deformation changes the shape of a body without changing its volume, since the total number of atoms and the crystal structure remain the same. A change in shape that occurs at constant volume can always be represented geometrically as the sum of elementary deformations of a type known as simple shear. A simple shear is the kind of deformation that deforms a cube into a parallelogram; it changes the angles between initially perpendicular directions in the cube. The shear due to the passage of an edge dislocation is illustrated in figure. While the dislocation translates the top of the crystal rigidly over the bottom to create a discrete step, the Burgers' vector has atomic dimensions, so the step is invisible. Macroscopic deformation is the sum of the slip caused by many dislocations.

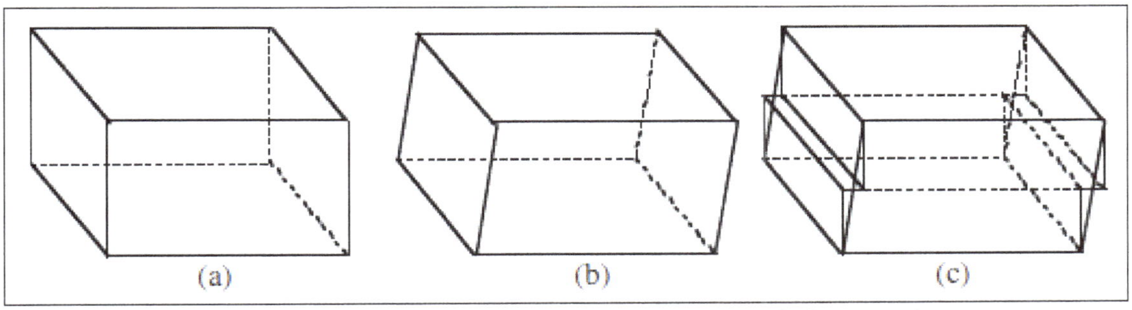

(b) is obtained from (a) by a simple shear of the top over the bottom. Figure (c) shows how the same shear can be caused by an edge dislocation

An Edge Dislocation in a Simple Cubic Crystal

The procedure that was used to create the edge dislocation that appears in figure made no reference to the structure of the solid, and can be used to form an edge dislocation in any material. However, when the material is crystalline the ordered pattern of atoms restricts the values that the Burgers vector, b, can have. The restriction is introduced by the welding step that is used to change the configuration. The welding must be so perfect that it is impossible to tell that the two surfaces were ever separated. If the solid is crystalline this can only be true if the crystal structure is continuous across the slip plane after the weld is made. It follows that the relative displacement across the slip plane must equal a lattice vector so that atoms can re-bond without changing their local atomic configurations. Since the relative displacement is equal to the Burgers vector, b, of the dislocation, b must be a lattice vector. If the dislocation is an edge dislocation, b must also be perpendicular to the dislocation line.

The geometry of an edge dislocation is relatively easy to visualize when the crystal has a simple cubic crystal structure. The atomic configuration around the dislocation line is more complicated in real crystal structures, but it is not necessary to deal with that complexity to understand the behavior of dislocations at the level we shall need in this course. Whenever we need to consider the crystallography of the dislocation we shall assume that the crystal structure is simple cubic.

An edge dislocation in a simple cubic structure is drawn in figure, which shows both a two-dimensional view and a three-dimensional section along the dislocation line. The dislocation can be created by making a cut in the crystal on the dashed plane that terminates at the dislocation line, displacing the material above the cut plane to the left of the dislocation by one lattice spacing, and allowing the atoms to re-bond across the slip plane. This recipe recreates the simple cubic unit cell everywhere except on the dislocation line itself (ignoring the small elastic distortion of the cells that border the dislocation line). Hence the Burgers vector, b, of the dislocation that is drawn in the figure is b = a, where a is a vector along the edge of the cubic unit cell.

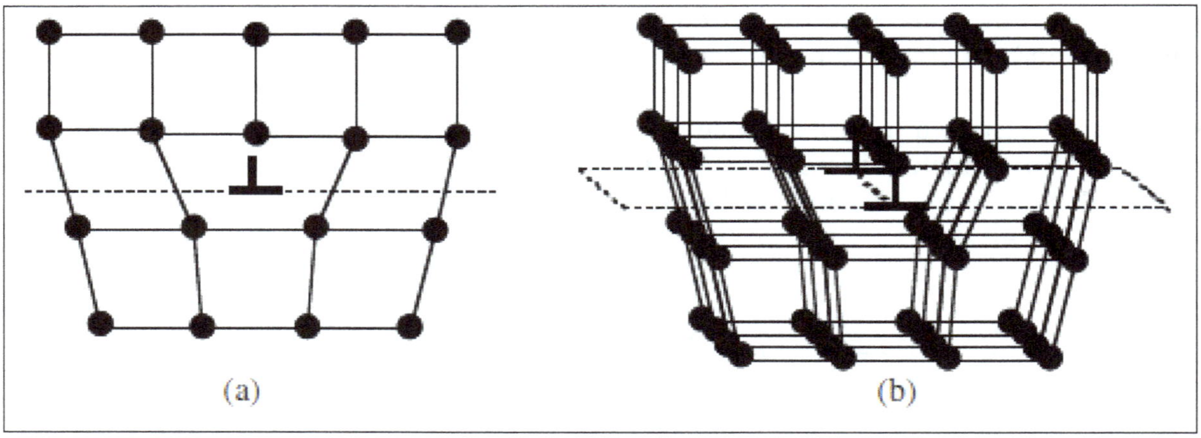

An edge dislocation in a simple cubic structure. The dotted plane is the slip plane.

The process that creates the edge dislocation shown in figure leaves one extra vertical half-plane of atoms above the slip plane. This extra half-plane terminates at the dislocation line, and is compressed there, as shown in the figure. The distortion at the dislocation line is local. The simple cubic arrangement of atoms is essentially restored a few atom spacings away from the dislocation line. The influence of the dislocation on the atomic configuration rapidly decays into a small displacement that decreases in magnitude with the inverse cube of the distance from the dislocation line. The local distortion near the dislocation line (or dislocation core) is indicated in the figure.

In principle, the Burgers vector of a crystal dislocation can be any lattice vector; for example, it is geometrically possible for an edge dislocation to be the termination of any integral number of lattice planes. In reality, however, the Burgers vector is almost invariably equal to the shortest lattice vector in the crystal. The reason is that the energy per unit length of dislocation line, which is called the line energy, or, in a slightly different context, the line tension of the dislocation, increases with the square of the magnitude of b, |b|2. (While we shall not prove this, it is obvious from figure that the local distortion of the crystal would increase dramatically if two or more extra half-planes terminated at the dislocation line.) Let the Burgers vector, b, be the vector sum of smaller lattice vectors, $b = b_1 + b_2$. Unless b_1 and b_2 are perpendicular, $|(b_1+b_2)^2 > |b_1|^2 + |b_2|^2$, and the dislocation can decrease its energy by splitting into two or more dislocations that have smaller Burgers vectors.

The Burgers Circuit

While it is always possible to find the Burgers vector, b, of a dislocation by determining the slip that would be required to make it, this is often inconvenient. A simpler method uses a geometric construction known as the Burgers circuit.

To construct the Burgers circuit, choose a direction for the dilocation line and draw a clockwise closed circuit in the perfect crystal by taking unit steps along the lattice vectors. An example is drawn for a {100} plane in a simple cubic crystal in figure. If the same circuit is drawn so that it encloses a dislocation, it fails to close. The vector (from the starting position) that is required to complete the circuit is the Burgers vector, b, of the dislocation, and measures the net displacement experienced by an imaginary observer who completes a circuit around the dislocation that would be closed in a perfect crystal.

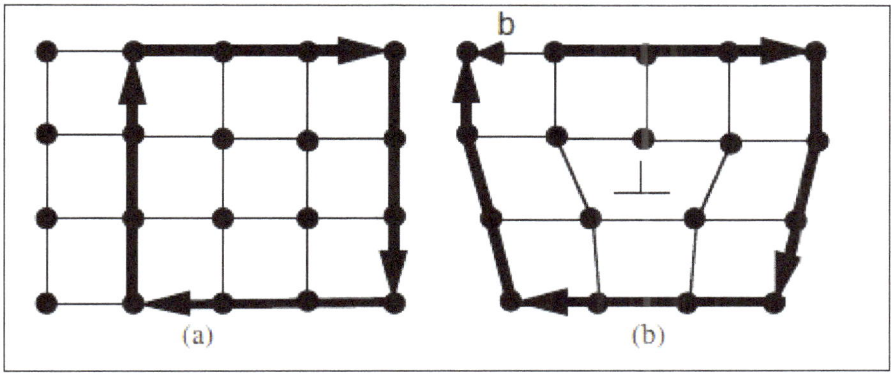

A Burgers circuit closes in a {100} plane of a cubic crystal, but fails to close by the Burgers vector, b, when the same circuit encloses an edge dislocation.

Motion of an Edge Dislocation: Glide and Climb

The reason that dislocations control the plastic deformation of crystalline solids is that it is relatively easy to move dislocations to produce shear deformation of the sort that is pictured in figure. It would be enormously difficult to shear a crystal by forcing the glide of rigid planes of atoms over one another; one would have to force the simultaneous reconfiguration of every crystal bond that crossed the slip plane. The same result is more easily achieved by moving dislocations stepwise through the crystal. Stepwise dislocation motion requires a much smaller force since each elementary step can be accomplished by reconfiguring only the bonds that neighbor the dislocation line.

The stepwise motion of an edge dislocation in a simple cubic crystal is illustrated in figure. In order for the dislocation to move one lattice spacing to the right it is only necessary to break the bond indicated by the long dash in figure and establish the bond indicated by the short dash. The new configuration is shown in figure. Of course one bond must be broken for each plane through which the dislocation threads, so a significant force is still required. But the force is small compared to that required to slip the upper part of the crystal as a rigid body. If the dislocation moves through the crystal in a sequence of individual steps like that shown in figure it causes a net slip of the material above its plane of motion by the Burgers vector, b, and hence causes a rigid displacement of the whole upper part of the crystal.

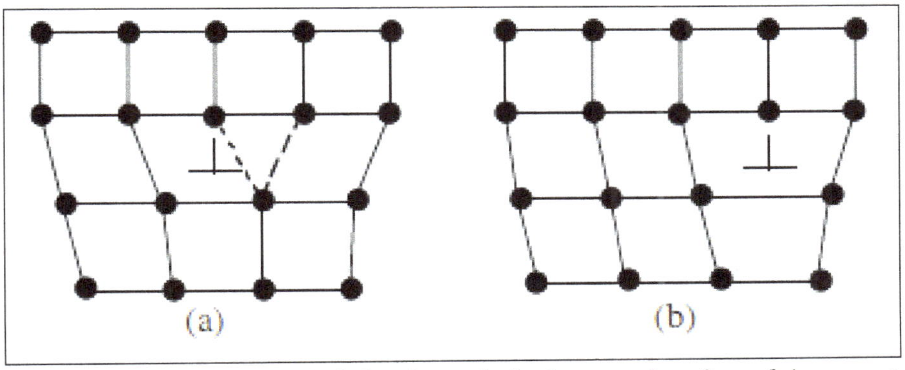

Glide of an edge dislocation. Only a single bond must be broken per plane for each increment of glide.

The type of motion that is illustrated in figure is called dislocation glide, and is relatively easy to accomplish. However, an edge dislocation cannot glide in an arbitrary direction. It can only glide in a particular plane, the slip plane or glide plane, which contains both the Burgers vector and the dislocation line.

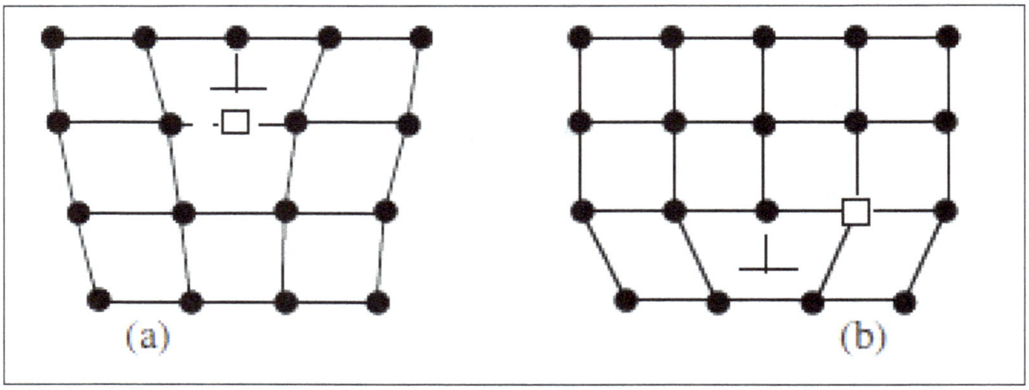

Climb of an edge dislocation. Movement up out of the plane requires the elimination of atoms by vacancies. Movement down requires the addition of atoms.

When an edge dislocation moves out of its glide plane its motion is called climb. The climb of a dislocation is difficult at ordinary temperatures since it requires that atoms be absorbed on or liberated from the extra half-plane of atoms that defines the dislocation line. The climb of an edge dislocation is illustrated in figure. The mechanism is slightly different depending on whether the dislocation moves up, which contracts the extra half-plane, or down, which extends it.

If the dislocation climbs up atoms must be liberated from the edge of the extra half-plane. Since the number of atoms is conserved, this requires the absorption of vacancies from the lattice. One vacancy is needed per plane the dislocation threads. If the dislocation climbs down it must add atoms to the extra half-plane, and can only do this by liberating one vacancy per plane into the matrix, as shown in figure. Both processes are difficult except at high temperature when, as we shall see, the equilibrium concentration of vacancies is high and the exchange of vacancies and atoms is relatively easy.

Because of the difficulty of climb at ordinary temperatures the plastic deformation of real crystals tends to occur through the motion of dislocations on well-defined planes that are the glide planes of the active dislocations. Under a microscope one can often see discrete slip steps on the surface of a crystal that has been deformed. These result from the glide of many dislocations on closely spaced, parallel planes. At high temperature climb becomes possible and the slip planes are less well-defined. When this happens the strength of the crystal (its resistance to plastic deformation) decreases dramatically. For this reason most solids are relatively soft at high temperature.

Screw Dislocations

Our discussion to this point has dealt only with edge dislocations, that is, dislocations in which the Burgers vector is perpendicular to the dislocation line. Dislocations in real crystals rarely have a pure edge character. Their Burgers vectors lie at various angles to their line directions. In the extreme case the Burgers vector is parallel to the dislocation line, which is the characteristic of a screw dislocation. A screw dislocation is difficult to visualize in a crystal, but can be created by a method suggested by Volterra that closely resembles the way the edge dislocation was formed. A screw dislocation of the general Volterra type is shown in figure.

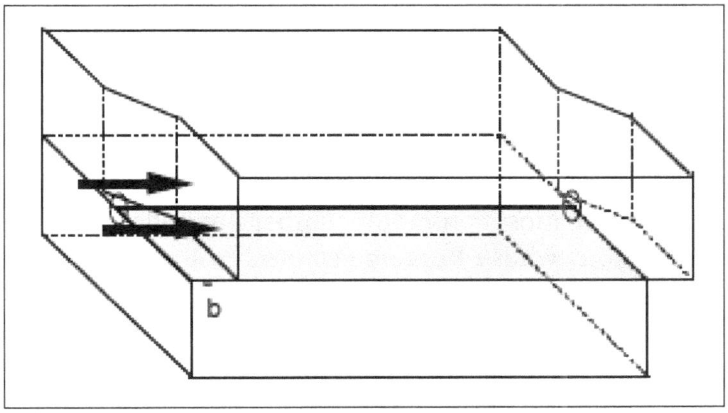

A method for forming a screw dislocation in a solid.

To introduce a screw dislocation we slice the solid part-way through in the direction of its width, as shown in figure. But instead of displacing the material above the cut toward the dislocation line we displace it by a vector, b, which lies parallel to the dislocation line, as shown in the figure. The direction of the force required to do this is also indicated. The material is then re-welded so that it is continuous across the plane of slip. The residual distortion is concentrated at the dislocation line, which then constitutes an isolated linear defect.

If the body shown in figure is crystalline then the cut surface shown in the figure is a plane of atoms. In order for the crystal to be continuous across the slip plane after it is rejoined, the displacement, b, must be such that this plane of atoms joins continuously onto a crystallographically identical plane. It follows that a closed circuit (Burgers circuit) that encloses a screw dislocation not only fails to close, but produces a net translation by b along the dislocation line, where b is a lattice vector. A circuit that starts on one plane of atoms finishes on another a distance b below. Continuing the circuit causes a displacement by b at each revolution, without the circuit ever leaving the atom plane. The effect of a screw dislocation is to join a set of parallel atom planes so that they become a single plane like one that would be created by extending a plane outward from the thread of a screw. Hence the name: screw dislocation.

As in the case of an edge dislocation the line energy of a screw dislocation is proportional to the square of its Burgers vector. Hence the Burgers vector of the screw dislocation is ordinarily the smallest lattice vector that is compatible with the direction of its line.

Screw Dislocations and Plastic Deformation

A screw dislocation differs from an edge not only in its geometry but in the way it accomplishes plastic deformation. The most important qualitative differences concern its direction of motion under an applied force and its relative freedom of movement.

Figure suggests the connection between slip and dislocation motion for a screw dislocation. As the screw dislocation is displaced through the width of the body the material above its plane is slipped in the direction of the Burgers vector, hence along the length of the body. It follows that the longitudinal force shown in the figure acts to drive the screw dislocation sideways. If a screw dislocation is passed through the full width of the body it causes the shear shown in figure, which the same as that is caused by the passage of an equivalent edge dislocation through the length.

In contrast to an edge dislocation, a screw dislocation can glide in any plane. Since the Burgers vector lies parallel to the dislocation line both are in any plane that contains the dislocation line and the screw dislocation can move in any direction perpendicular to its line.

Dislocations in Real Materials: Mixed Dislocations

Dislocations in real materials are most commonly neither pure edge nor pure screw in their character, but are mixed dislocations whose Burgers vectors lie at an intermediate angle to the local direction of the dislocation line. Because of the way in which they interact with other elements of the microstructure (which we shall discuss at a later point) dislocation lines are ordinarily curved. Since the dislocation bounds a region that has been slipped by the Burgers vector, b, the Burgers vector is the same at every point on the dislocation line. Hence the character of a curved dislocation changes continuously along its length. For this reason it is often most useful to think of a dislocation as the boundary of a surface over which the crystal has slipped rather than as a defect with a particular local atomic configuration.

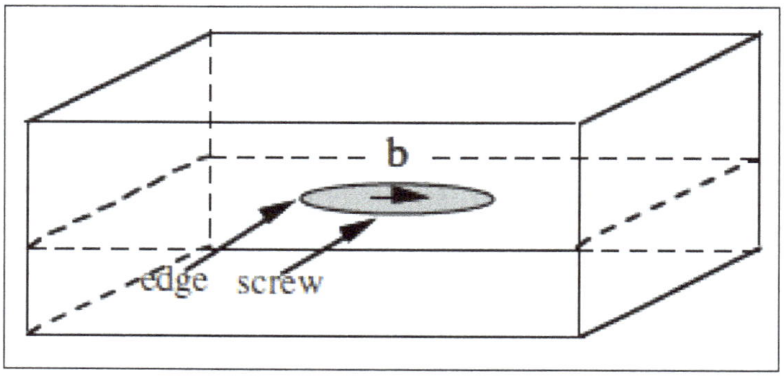

A dislocation loop in a crystal.

This point is illustrated by the dislocation loop shown in figure. Dislocation loops are relatively common features of the microstructures of structural alloys. The loop shown in the figure is created by a process that is equivalent to a slip of the cylinder of material above the loop by the vector b. The dislocation loop is the boundary of the surface on which slip has occurred. The character of the dislocation changes continuously around the loop. The dislocation has pure edge character at the extremities of the loop along the length of the body, and has pure screw character at the extremities along the width.

The qualitative nature of the force on a curved dislocation, such as the loop in figure, can also be understood by regarding it as the boundary of the slipped region. If the force on the body is oriented to shear the body by the vector b then the dislocation will end to move so that it increases the area of slip; the loop will expand. If the force acts to shear the body in the direction -b, the dislocation will tend to move so that it shrinks the slipped area; the loop will contract.

Whether edge, screw or mixed in character, all dislocations have three important properties. First, they bound surfaces that divide the crystal so that the part above the surface is slipped by the vector, b, with respect to that below. Second, the slip on the surface bounded by the dislocation is constant and equal to a lattice vector. Hence the Burgers vector, b, is a lattice vector and has the same value at every point on the dislocation line. Third, since the slip is constant, a dislocation cannot simply end within a material; there is always a boundary between a slipped and an unslipped area.

The dislocation can terminate at a free surface, it can close on itself, as in figure, or it can end at a junction with other dislocations that bound surfaces over which the magnitude or direction of slip is different.

Partial Dislocations

In the original concept of Volterra, the Burgers' vector of a crystal dislocation is a translation vector of the crystal, that is, a vector that connects atom positions so that the crystal can be rejoined perfectly across its glide plane. Such a dislocation is called a total dislocation. The element of slip it induces carries an atom in the glide plane from one atom position to another. Only total dislocations can be true two-dimensional defects. If the Burgers' vector were not a translation vector of the crystal the slipped surfaces would not weld perfectly together across the glide plane and a planar, surface defect would be left behind whenever the dislocation moved.

While most crystal dislocations are total dislocations when viewed from sufficiently far away, it is not uncommon to find them dissociated locally into a configuration that can be described as two parallel partial dislocations connected by a planar defect that is called a stacking fault in the crystal. The prototypic example is found in FCC crystals.

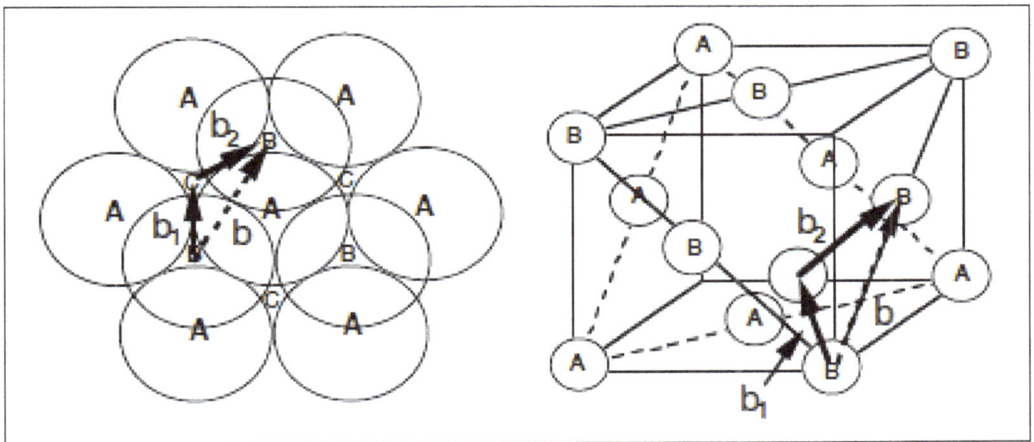

Illustration of the slip of close-packed planes. The dashed arrow shows slip by a total dislocation, $b=\frac{1}{2}[\bar{1}01]$, which carries $B \rightarrow B$. The solid arrows show the competing process of successive slip by two partial dislocations. The shaded atom in the cell is the intermediate, C-site position reached after partial slip.

The common dislocation in the FCC structure is the dislocation that causes close-packed planes to slip over one another. The close-packed planes are {111} planes, and the Burgers' vectors of the dislocations that induce slip on these planes are the $\frac{1}{2}<110>$ vectors that connect atoms to their nearest neighbors in the {111} planes. Figure illustrates the path by which an FCC crystal is slipped by one atom spacing in the (111) plane. The slip vector is $b=\frac{1}{2}[\bar{1}01]$, and is the slip caused by a dislocation with Burgers' vector $b=\frac{1}{2}[\bar{1}01]$. Assuming that the glide plane of the dislocation lies between (111) planes of A and B-type atom sites, the element of slip carries an atom from one B-site to another, as shown in the figure. For clarity, the figure is drawn twice. The drawing on the left shows the displacement in a stacking of close-packed planes (A,B). The drawing on the right places the A and B planes in an FCC unit cell.

Direct slip from one B position to another is relatively difficult, as one can easily show by trying to slide two close-packed arrays of balls over one another. The atoms (balls) in the B-plane must ride over the A atoms, as illustrated in the figure. It is much easier to accomplish the slip in two sequential steps, as indicated in the figure, that circumvent the A atoms. The B atoms are first slipped into C positions, then moved from C back to B again. The slip can be accomplished by the sequential passage of two dislocations with the Burgers' vectors b1 and b2. However, b_1 and b_2 are not lattice vectors; they are examples of partial dislocations. Their sum is the total dislocation, b.

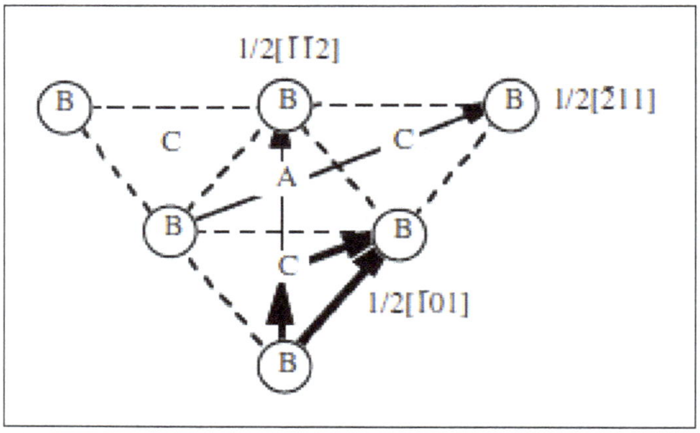

Configuration of atoms in the (111) plane showing the partial and total slip vectors.

To find the Burgers' vectors of the partial dislocations let the close-packed plane be the (111) plane, as shown in the figure. The atom shift is to the shaded atom position, followed by a shift to the final atom position in the face center. The (111) plane of B-sites in the unit cell in figure is redrawn in figure. As shown in this figure, the Burgers' vector of the total dislocation is $b=\frac{1}{2}[\bar{1}01]$. The first partial slip is from the B site at the corner of the unit cell to the nearest C-site in the plane. This slip is 1/3 of the vector connecting B-sites along the height of the triangular configuration of atoms in the plane, which is $b=\frac{1}{2}[\bar{1}\bar{1}2]$. Hence,

$$b_1=\frac{1}{6}[\bar{1}\bar{1}2]$$

The second partial slip is from the C-site to the final B-site position. As shown in the figure, this vector is parallel to the vector $\frac{1}{2}[\bar{2}11]$, and is 1/3 of its length. Hence,

$$b_2=\frac{1}{6}[\bar{2}11]$$

The vector sum of the two partial dislocation vectors is:

$$b_1+b_2=\frac{1}{6}[\bar{1}\bar{1}2]+\frac{1}{6}[\bar{2}11]=\frac{1}{2}[\bar{1}01]=b$$

Splitting the total dislocation, b, into the partials, b_1 and b_2, not only facilitates slip, but also lowers the total energy. The energy per unit length of dislocation line is proportional to $|b|^2$. Since,

$$|b|^2 > |b_1|^2 + |b_2|^2$$

The energy of a parallel pair of partial dislocations, b_1 and b_2, is less than that of the total dislocation, $b = b_1 + b_2$.

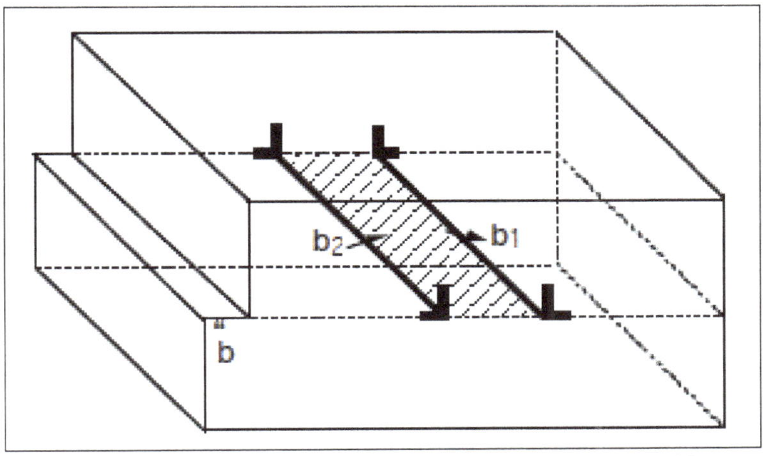

The total dislocation, b, divided into partial dislocations, b_1 and b_2 separated by a stacking fault.

However, to understand the configuration that appears when the dislocation is split we must recognize that eq. $|b|^2 > |b_1|^2 + |b_2|^2$ gives only part of the energy of the pair of partials. Let the total dislocation divide, as illustrated in figure. In the area between the partials, which is shaded in the figure, the region above the slip plane has been slipped by the vector b1, but has not yet experiences the slip b_2. Let the slip plane lie between A and B-planes, as illustrated in figure. The partial slip b_1 carries the atoms in B-sites into C-sites, and causes corresponding changes $C \rightarrow A$ and $A \rightarrow B$ in successively higher planes. The FCC crystal has the stacking ..ABCABC.. But the stacking encountered in a direction perpendicular to the glide plane that passes through the shaded area between the partials is ..ABCA|CABC.., where the slash indicates the glide plane. The partial dislocation perturbs the sequence of close-packed planes; it introduces a planar defect called a stacking fault on the glide plane between the two partials.

A stacking fault has a positive energy, σ_s, per unit area. The separation of the two partials, b_1 and b_2, creates a stacking fault whose area increases as the partials move apart. There is, therefore, an equilibrium separation between the partial dislocations. They separate to decrease the energy per length of dislocation, but are attracted by the need to minimize the stacking fault energy. The balance of these two effects determines the equilibrium separation, which is of the order of 5-500 A in typical FCC crystals.

Similar considerations apply to HCP and diamond cubic crystals. Total dislocations in the close-packed planes tend to divide into partials separated by ribbons of stacking fault. In BCC crystals the division of total dislocations into partials is much less pronounced. Separated partials cannot ordinarily be resolved in BCC, even though high resolution electron microscopy. Their presence is inferred from features of the mechanical behavior that are most easily explained by assuming that total dislocations dissociate slightly.

Surface Defects

Surface defects are defects of two dimensional natures in the mathematical sense (that is, they have two dimensions). These are areas of distortions that lie about a surface, having a thickness of a few atomic diameters (that is, the thickness is almost negligible compared to other two dimensions of the surface). From the thermodynamics point of view, surface defects are not stable-and are present as metastable defects, and under suitable conditions, these defects can be removed.

Types of Surface Defects

Free Surfaces

Free surfaces (external surfaces) of materials are surface defects in the sense that growth of crystal has stopped abruptly at the surface. The atoms in the free surface of a solid (or, even liquid) have no neighbours and have no cohesive-bonds, on one side.

That is why, these atoms have higher free energy as compared to atoms inside. Free surfaces have surface energies that are roughly calculated from the number of 'broken bonds' at the surface. For example if the free surface is a close-packed plane of a close-packed crystal structure such as {111} type.

An atom in free surface has six neighbouring bonds in the surface plane, and three bonds with atoms below this surface but the three bonds above the surface are broken (as coordination number is 12). Thus, three out of total twelve bonds of an atom, that is, one quarter of bonds are missing compared to an atom inside.

The surface energy shall be one-quarter of the cohesive energy per atom, that is, around 1 eV for a metal such as FCC-iron. There are on an average around 10^{15} atoms cm^{-2} in a crystal surface. So, the surface energy becomes 10^{15} eV cm^{-2} i.e., 1600 ergs cm^{-2}. Table shows experimentally measured surface free energy of some metals.

Table: Surface free energy of some materials.

Metal	Ag	Au	Cu	Fe	Ni	Pt	Sn	Si	NaCl	LiF
γ, J/m^2	1.13	1.35	1.73	1.40	1.73	2.10	0.69	1.24	0.30	0.34

Free surfaces can be classified as (a) Singular surface, (b) Vicinal surface as illustrated in figure.

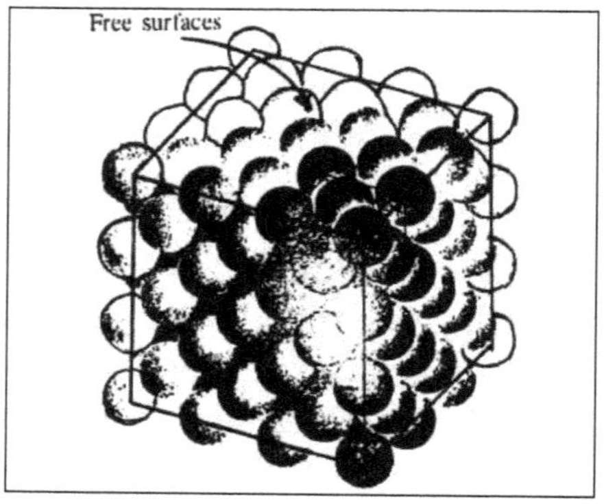

Schematic structure of free surface of a block of crystal.

Singular surface has low surface energy and usually have low index planes such as (100), (110), or (111). Vicinal surfaces have orientation near those of singular surfaces and are formed of low index planes separated by mono-atomic or mono-molecular steps or ledges, kinks and double kinks, etc.

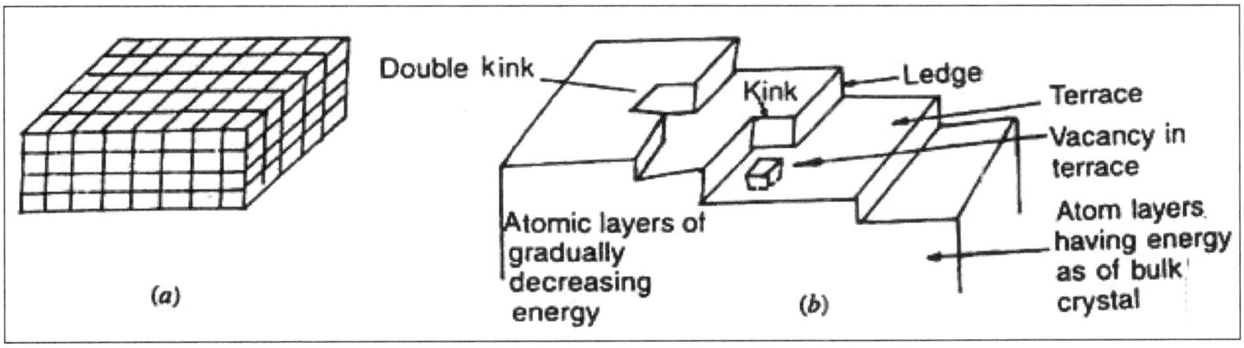

Type of free surfaces (a) Singular Surface, (b) Vicinal Surface.

In metals, the free (valence) electron bond is unselective and that is why, any two metals with clean surfaces, when brought in close contact adhere well. And, it follows that a liquid metal of low cohesive energy generally spreads freely over a clean surface of a metal of high cohesive energy, because by doing so, a surface of high energy (solid metal) is replaced by a low-energy surface and a low-energy metal-metal interface. The wetting action of solders and brazing alloys is based on this principle.

Grain Boundaries

Classification of Boundaries

Dislocation theory offers good means for broad understanding of the nature and properties of grain boundaries. In a single phase polycrystalline material, a grain boundary is a two dimensional surface, which separates two regions differing in orientation of their crystal axes.

It is a surface defect where orientation of unit cells of one grain changes to other, i.e. at this surface; there is a mismatch of unit cells with unlike orientations. There are many types and gradations of grain boundaries in metals and alloys.

They are classified as:

- Based on Degree of Difference in Orientation as:
    - Sub-Boundaries: These are boundaries supposed to be existing within a grain, when the two regions on both sides of them have as small a difference in orientation as 1°, or less.
    - Low-Angle Boundaries: Low-Angle Boundaries are boundaries between two adjacent grains differing in orientation by less than about 10°.
    - High Angle Boundaries: High angle boundaries are common boundaries in materials having large difference in orientation between grains.

- Based on Type of Difference in Orientation as:
  - Tilt Boundaries: Tilt boundaries are boundaries between adjacent grains having misorientation in such a way that lattices appear to be tilted with respect to one another.
  - Twist Boundaries: Twist boundaries are boundaries between adjacent grains having misorientation in such a way that the lattices appear to be rotated (twisted) with respect to one another.
  - Coincidence Boundaries: Coincidence boundaries are boundaries in which atomic positions common to both the adjacent grains form some fraction of the total boundary area.
- Based on Continuity of Atoms on both Sides of the Boundary:
  - Coherent Boundary: Coherent boundary is a boundary where there is complete continuity of atoms across it, i.e. there is one to one correspondence of atoms at the boundary/interface.
  - Semi-Coherent Boundary: Semi-coherent boundary is a boundary where there is partial continuity of atoms across it, i.e., some planes of atoms are continuous, but some may end at it.
  - Incoherent Boundary: Incoherent boundary is a boundary having no coherency of atoms across it at all.

Sub-boundaries

When a single crystal is bent by cold working, and if one slip system operates (More dislocations are generated on one slip system), then the bent crystal looks as in figure, because excess positive edge dislocations remain in the crystal to accommodate the curved shape of the crystal.

When such a bent crystal is annealed (before the recrystallisation takes place), the curved crystal forms a number of strain-free sub-grains, each preserving the local orientation of the original bent crystal separated by plane sub-boundaries. This process of forming sub-grains and sub-boundaries is called polygonisation, because a smooth curved crystal changes into polygons.

The dislocations have arranged themselves in sub-boundaries (by the processes of slip and climb during annealing), one above the other because in this position, dislocations relieve each other's elastic strain fields. The dislocation arrangement is one of high strain energy, which changes to as in figure of lower strain energy.

The sub-boundaries between the sub-grains are considered to exist within a grain and thus, the difference of orientations between the two adjacent sub-grains is very small, around 1° or less. Sub-boundaries are small-angle or symmetric tilt-boundaries within the grain having very small difference of orientation between the sub-grains. Sub-boundaries are revealed by etching as dense row of pits under microscope.

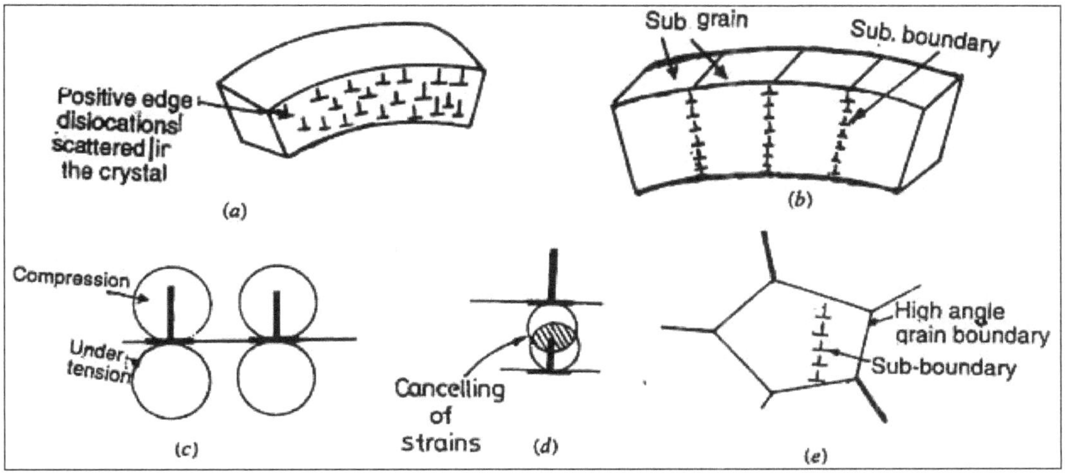

shows formation of sub-boundaries. (a)A bent cold worked single crystal with excess dislocations of one sign (here positive edge dislocations), (b) After polygonisation, sub-boundaries are formed within the crystal itself, (c) Dislocation on same plane, (d) Tensile strain of upper dislocation is partly cancelled by compressive strains of lower dislocation, (e) Schematic representation of array of parallel positive edge dislocation within a grain at a sub-boundary.

Low-angle Boundaries or Tilt-boundaries

In 1940, both Bragg and Burger, and later Read-Shockley proposed the dislocation model of structure of low-angle boundaries with good success. Low angle boundaries are boundaries between adjacent crystals of same structure but with small difference in crystal orientations, and might be considered to have an array of parallel dislocations (along the plane of the boundary).

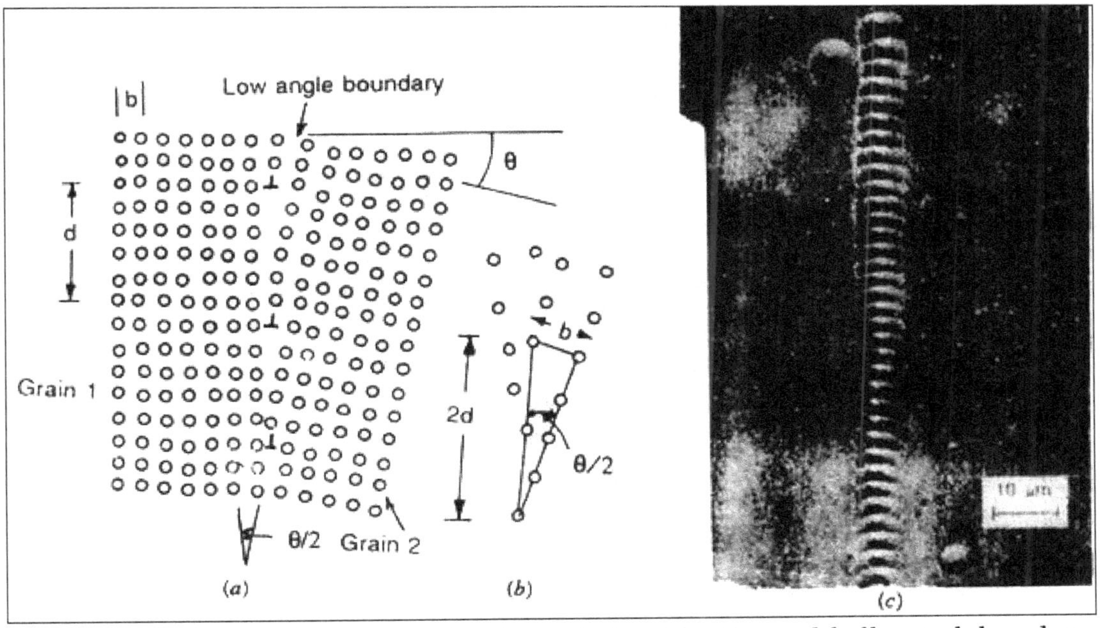

Nature of low-angle boundary (tilt boundary) in metals. (a) Dislocation model of low-angle boundary, (b) Relationship between θ, the angle of tilt with d, the spacing between two dilocations, (c) Etch pits seen under microscope revealing the low angle boundary having regularly spaced dislocations.

Two grains in schematic are misoriented or pictured as tilted symmetrically about an axis (horizontal inside the paper), parallel to the direction of edge dislocations. The two lattices can be maintained continuous by periodic addition of an edge dislocation.

The resulting vertical array of parallel dislocations of same sign (here, positial edge dislocations) is just like a sub-boundary but with greater difference of orientation, still less than about 10°, and is in between two adjacent grains. Low-angle boundaries are also observed under microscope as uniformly spaced line of etch pits.

The presence of each dislocation at the surface of a specimen as etch pit is revealed because of the faster rate of chemical attack by the etchant in the elastically strained area of the crystal near the core of the dislocation.

The angle theta is the angle of misorientation. The greater is the angle of misorientation, the greater is the inclination of planes that terminate as dislocations at the boundary, and closer is the spacing of the dislocations in the vertical boundary, i.e., the spacing between the dislocations in the boundary determines the theta, the angular inclination between the lattices. In figure, we find that:

$$\sin\theta/2 = b/2d$$

Where, b is the burgers vector of dislocation in the boundary, and d is the spacing between the two adjacent dislocations. If the angle of misorientation is assumed to be small, then $\sin\theta/2 \approx \theta/2$, and

$$\theta = \frac{b}{d}$$

And the result of this relationship has been found to be in good agreement by x-ray diffraction determinations, and confirms the validity of the dislocation model for low-angle boundaries.

This relationship could be used to drive the energy of the boundary as dependent on angle of misorientation. With the increase in misorientation angle, θ, of a symmetric tilt boundary, the value of d decreases, i.e., the number of dislocations increases.

We know by now that total energy of each edge dislocation per unit length is given by:

$$E_b = E_C + \frac{Gb^2}{4\pi(1-v)} \text{In} \frac{d\cdot}{2r_0}$$

Where, E_c is the energy per unit length of dislocation core. Energy of the dislocation decreases as d becomes small, or θ increases (∵ d = b/θ). If we assume r_o = 2b, the total energy of the dislocation per unit length,

$$E_b = E_C + \frac{Gb^2}{4\pi(1-v)} \text{In} \frac{1}{4\theta}$$

If 'd' is distance between two dislocations, then energy of unit area of the boundary is approximated as:

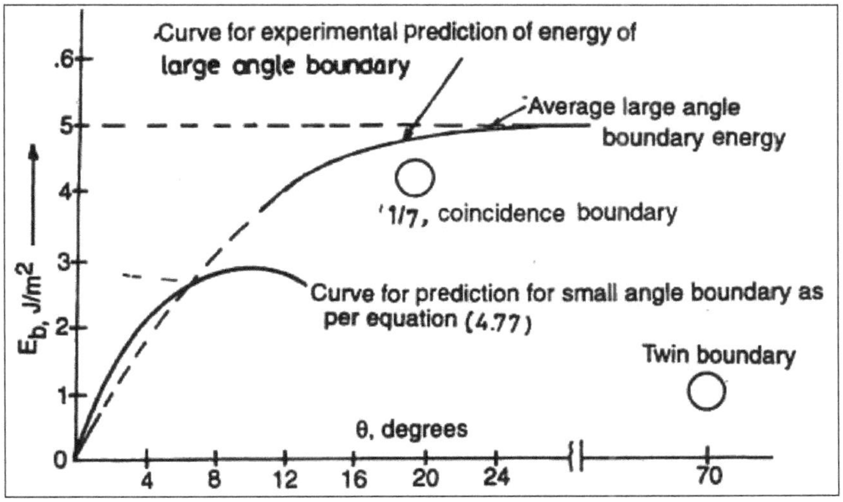

Comparison of energy for experimental data and theoretical data for copper.

$$E_b \approx \left[E_C + \frac{Gb^2}{4\pi(1-v)} \text{In} \frac{1}{4\theta} \right]$$

$$= \frac{\theta}{b} \left[E_C + \frac{Gb^2}{4\pi(1-v)} \text{In} \frac{1}{4\theta} \right] = \theta \left[\frac{E_C}{b} + \frac{Gb}{4\pi(1-v)} \text{In} \frac{1}{4\theta} \right]$$

$$= \theta \left[\frac{E_C}{b} - \frac{Gb}{4\pi(1-v)} \text{In}\, 4 - \frac{Gb}{4\pi(1-v)} \text{In}\, \theta \right]$$

$$\approx \theta [A - B \text{ In}\, \theta]$$

Where,

$$A = \frac{E_C}{b} - B\, \text{In}\, 4, \text{ and}$$

$$B = \frac{Gb}{4\pi(1-v)}$$

If a graph is plotted between E_b and θ as in figure and experimental data is superimposed for copper, the curve matches well up to below about 8°. There is no maximum obtained experimentally. The equation $\theta[A - B \text{ In}\, \theta]$ predicts well that as b and E_C approach zero, the energy vanishes.

At higher angles than 8°, the dislocations are quite closely spaced, and are difficult to be distinguished due to overlap of cores of the dislocations. The structure of the grain boundaries beyond 8° of misorientation appears to acquire more complicated configurations. This model of dislocations appears to be successful for low angle boundaries up to about 8°.

Problem: Copper shows array of edge dislocation an (110) plane of atoms. Calculate the spacing of dislocations if the angle of tilt is 1°. (Lattice constant of copper 3.62 A).

Solution: Copper is FCC, and its (110) plane is shown on which dislocations have Burgers vector,

$$\therefore \text{Burger Vector} = \frac{3.62}{\sqrt{2}}$$

$$= 2.56A$$

As,

$$\theta = 1° = \frac{1 \times \pi}{180} \text{radians.}$$

$$\therefore d = \text{distance between two neighbouring dislocations}$$

$$= 2.56 \times \frac{180}{\pi}$$

$$= 146.70\,A°$$

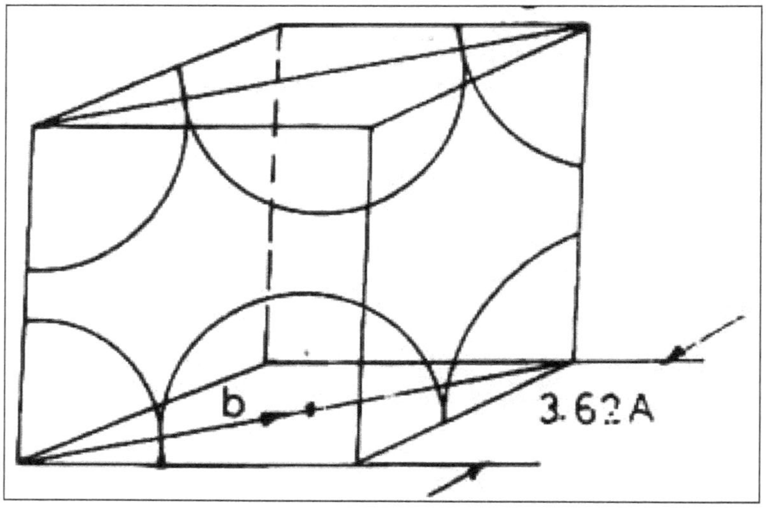

Degrees of Freedom of Grain Boundary

All single phase poly-crystalline materials contain a complex network of grain boundaries between adjacent grains that are crystallographically misoriented with respect to one another. The grain boundary described in figure, where the two neighbouring grains appear to be inclined or tilted with each other (by a small angle) about a horizontal axis normal to the paper is a very special boundary, as it has only one degree of freedom, or variable. In fact, the grain boundaries have five degrees of freedom, three to define the rotation, and two to define the boundary plane.

We specify the difference in orientation by the rotations which we would have to give to one crystal to bring it into same orientation as the other crystal. For example, if we rotate the grain 1 in figure by an angle theta clockwise about a horizontal axis which runs inside normal to the paper, the two grains become in the same orientation.

There are thus, three degrees of freedom available for such rotations, i.e. there are three ways of doing it corresponding to the three perpendicular axes about which rotations can be made. There are two more degrees of freedom for choosing the orientation of the grain boundary itself, that is, the boundary may not be in symmetrical position between the two crystals, but can be rotated about either of the two axes at 90° to each other.

We can tilt or twist one crystal relative to the other in three ways, and have two ways to align the boundary between the crystals. Figure illustrates schematically these five degrees of freedom of a grain boundary. In figure, the two neighbouring crystals are tilted with respect to each other about a horizontal axis, normal to the plane of paper with grain boundary placed symmetrically.

This boundary has one degree of freedom and is a symmetrical tilt boundary. Figure illustrates also a symmetric tilt boundary but with a vertical tilt axis. In both these cases, the rotating tilt axis lies in the boundary. The boundary has edge dislocations in it. In Figure, one crystal has been rotated relative to the other about a horizontal axis perpendicular to the boundary.

This is called a symmetric twist boundary, and normally contains at least, two different arrays of screw dislocations. Figure illustrate that the grain boundaries do not have to be in symmetric positions between the two crystals, but may align asymmetrically in two different ways between the crystals.

Majority of the grain boundaries in single phase polycrystalline materials have a mixed character of these five variables of varying degree, and thus the general or large angle boundaries have a complex structure.

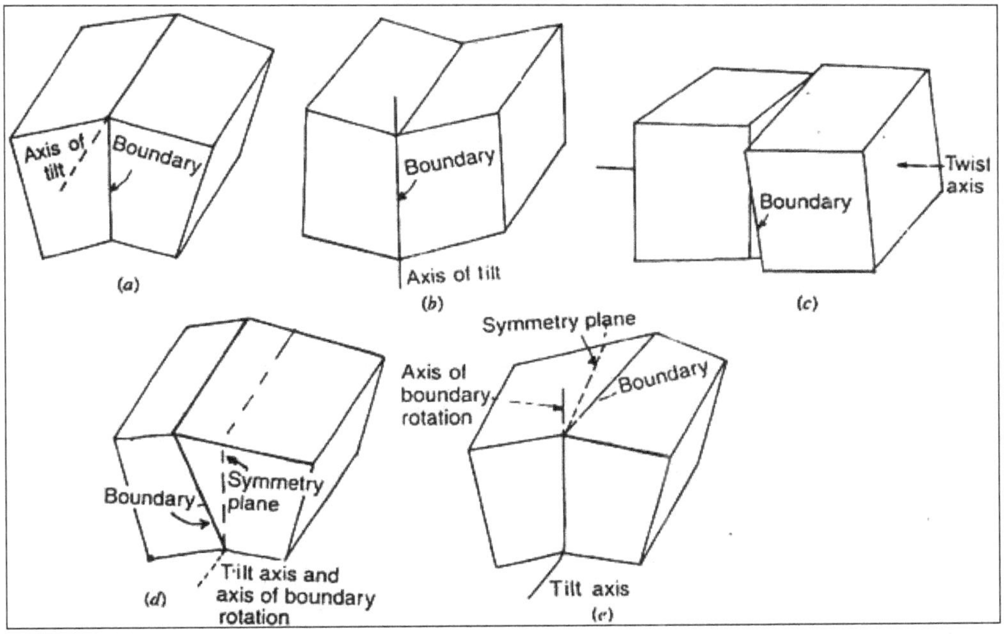

Five degrees of freedom of a grain boundary. (a)Symmetric tilt boundary with horizontal axis of tilt, (b) Symmetric tilt boundary with vertical tilt axis, (c) Symmetric twist boundary, (d) Asymmetric tilt boundary, (e) Asymmetric tilt boundary with vertical axis of boundary rotation.

A symmetric low angle boundary, or symmetric tilt boundary, figure has angle of misorientation equally divided between the two neighbouring grains, i.e., each lattice is displaced by $\theta/2$ from the vertical axis, i.e., has one degree of freedom. But a typical tilt boundary may have two degrees of freedom (or even more), such as asymmetric tilt boundary, figure, where the angle of misorientation is not divided equally between the grain 1 and grain 2.

Let us take a special case, where the angle has been so chosen that grain 1 contributes twice as many edge dislocations of Burgers vector b1→, as by grain 2 of Burgers vector, b2→. Let us assume that 2 dislocations of grain 1 combine with one dislocation of grain 2, by reaction,

$$2\,\vec{b_1} + \vec{b_2} = \vec{b_3}$$

To give a resultant average dislocation of Burgers vector, although this reaction does not occur, but it can be assumed that these dislocations of Burgers vector, lie along the asymmetric tilt boundary, and is the Burgers vector perpendicular to such an asymmetric boundary.

Depending on the position of the asymmetric tilt boundaries, the angle θ of misorientation may be shared in innumerable ways between the grains to give an average resultant dislocation of different Burgers vectors. Thus, a large number of such asymmetric boundaries can be assumed to exist.

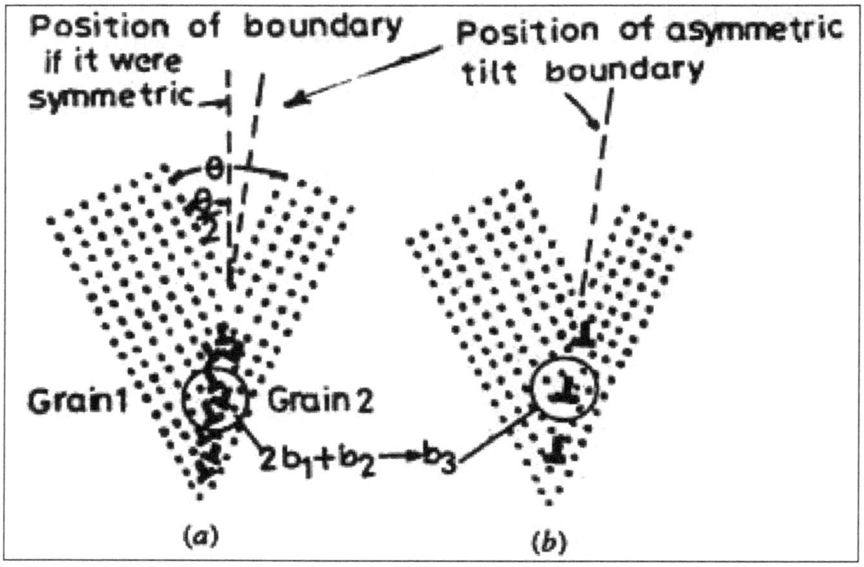

shows asymmetric tilt boundary with two degrees of freedom. (a) Asymmetric tilt boundary. As the boundary has been rotated along horizontal axis, gain 1 adds twice as many dislocations as grain 2. (b) Same as (a), but 2 dislocations of grain 1 and one of grain 1 and one of grain 2 result to give an equivalent average dislocation ($2\,\vec{b_1} + \vec{b_2} = \vec{b_3}$), b_3, which are assumed to be lying along the boundary.

Twist Boundary

A small-angle pure twist boundary is obtained, if one of the crystals appears to be rotated or twisted relative to its neighbour about an axis perpendicular to the plane of the boundary.

In figure, we see shape of cross section at the slip plane of a square crystal lattice deformed by a set of vertical uniformly spaced screw dislocations. The two surfaces on either side of the slip plane have been displaced by an angle, θ, but such a set of vertical dislocations have not accomplished the desired rotation needed to get a twist boundary of figure.

If only a horizontal set of screw dislocations are introduced, then the result is illustrated in figure, which also does not match with other figure. If simultaneous introduction of two sets of screw dislocations is done, we get rotation, or twist of two crystals relative to each other.

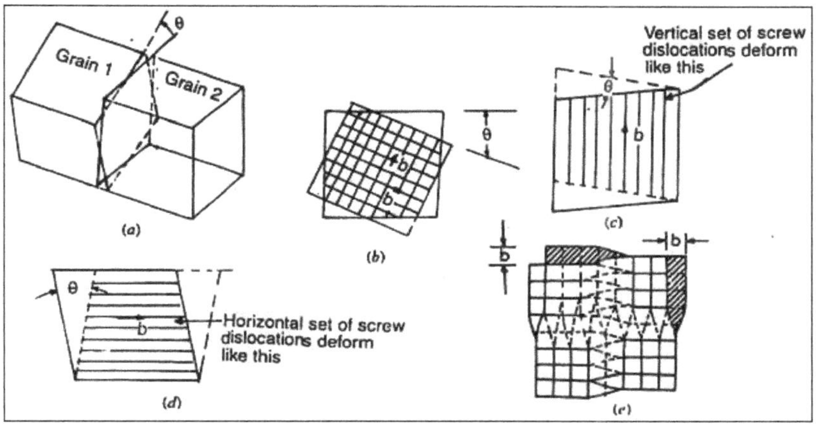

shows twist boundary. (a) Grain 2 has been twisted with respect to grain 1 along an axis perpendicular to boundary. (b) Both type of screw dislocations, that is, crossed grid of screw dislocations causes the twist to permit a transition between grain 1 and 2, (c) Shape of cross-section at slip plane of a square crystal deformed by a set of vertical screw dislocations, (d) by a set of horizontal screw dislocations, (e) Two intersecting screw dislocations of (d) and (c).

A twist boundary, lying along a crystallographic plane has rotational displacements of atoms symmetric with respect to the boundary. This type of boundary requires crossed grid of screw dislocations to permit continuity from grain 1 to grain 2. If in a unit area of boundary, each set of screw dislocations contain n dislocations, then the twist angle, θ, is:

$$\theta = n\vec{b}$$

If the twist boundary does not lie along a crystallographic plane, then, it is asymmetric with respect to rows and columns of atoms on either side of it. At least, then, three sets of screw dislocations with different Burgers vector are needed to produce an asymmetric twist boundary.

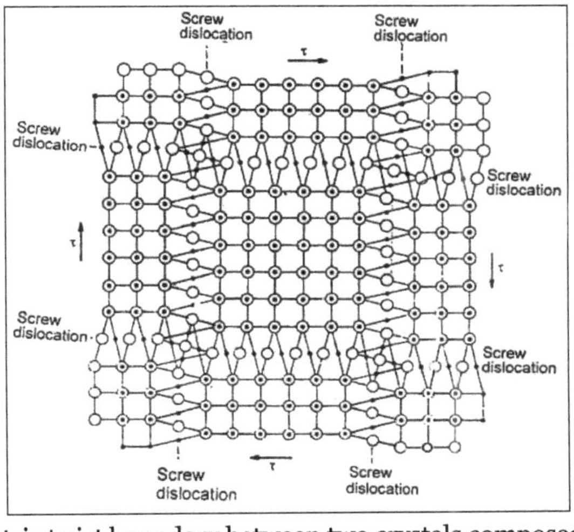

Symmetric twist boundary between two crystals composed of two sets of intersecting screw dislocations in a simple cubic lattice.

An asymmetric twist boundary is analogous to asymmetric tilt boundary. The actual small angle boundary of arbitrary orientation may have varying five degrees of freedom that is, varying degree of tilt and twist of crystallographic axis. If the boundary is asymmetric, then, it may, contain edge, screw and mixed dislocations.

Coincidence Boundary

If we make duplicate square lattice of holes in a black paper, and then pin the two sheets together, say at X in figure and hold them against light and rotate into positions in which various other points come into coincidence.

A rotation of 37° about a lattice point X, makes atom Y to coincide with Y', and Z to Z' in the un-rotated lattice. The figure illustrates two neighbouring grains in which grain 2 has been shown after rotation of 37° relative to grain 1. Dark dots are coincidence points.

It is clear that coincidence sites are continuous in both the grains in spite of grain boundary between them. These coincidence sites appear to form a lattice of themselves (joined by fine lines), but of larger cell dimensions. We see that one atom out of five atoms is a coincidence site, so that the term 1/5 coincidence boundary is applied to this arrangement of points.

Table illustrates some coincidence lattices in BCC and FCC metals. Coincidence lattices occur in all of the common crystal structures, and have site densities varying from 1/3 to 1/19 and even lower. Coincidence boundary normally is incoherent boundary like an ordinary grain boundary; that is, most of the atoms at the boundary of one crystal do not match the lattice positions in the other crystal, but mismatch is less, and so have less energy than ordinary boundaries.

Table: Some coincidence lattices in FCC and BCC Structures.

Structure	Axis of rotation	Angle of rotation	Density of coincidence sites
FCC		22.0	1/7
		38.2	1/7
		38.9	1/9
		36.9	1/5
BCC		60.0	1/3
		38.9	1/9
		70.5	1/3
		36.9	1/5

As coincidence sites are low energy sites, the grain boundary between the grains may not follow a relatively straight path but a zig-zag path of figure, so that the grain boundary has high density of coincidence sites. Most coincidence lattices do not have high densities of coincidence sites, but in certain cases, the coincidence sites lie in moderately close-packed planes, leading to the formation of coherent boundary. As coincidence sites are low energy sites, the grain boundary between the grains may not follow a relatively straight path but a zig-zag path of figure, so that the grain boundary has high density of coincidence sites. Most coincidence lattices do not have high densities of coincidence sites, but in certain cases, the coincidence sites lie in moderately close-packed planes, leading to the formation of coherent boundary.

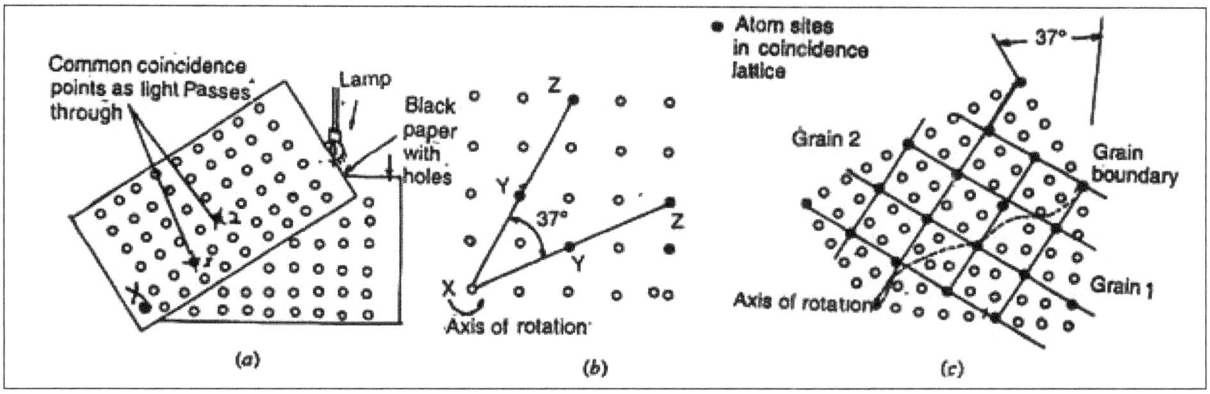

Figure shows coincidence lattice in square lattice (two dimensional). (a) Two black card-board with holes of square lattice and both pinned at X, and one is rotated keeping the other fixed and seen against a lamp. Two coincidence points, 1 and 2 are seen. (b) A square lattice when rotated about its origin X by 37° make atom Y to coincide with Y' and Z to Z'. (c) When two adjacent square lattice are misoriented by 37°. One atom out of five is a coincidence site, i.e. 1/5 coincidence lattice exists.

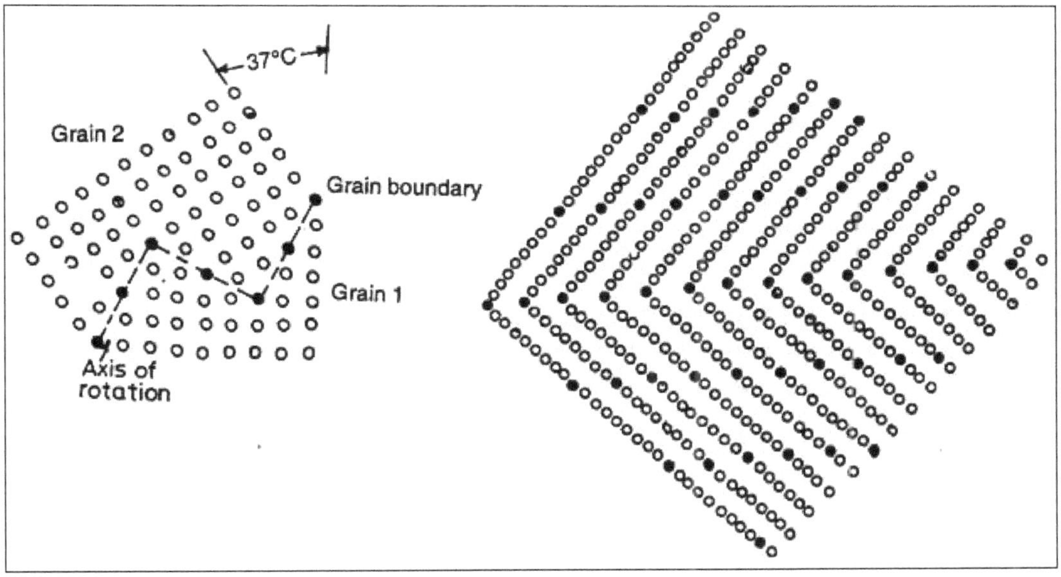

(a) Grain boundary takes a zig-zag path to pass as many coincidence sites as possible. (b) In BCC lattice, the neighbouring (210) planes obtained after a rotation of 94°-4'. A high density coincidence boundary which is also a coherent boundary.

Coherent, Semi-Coherent and Incoherent Boundaries

A coherent boundary has all the atom sites in it common to both crystals (regions) joined by it. There is, thus, continuity of crystal structure from one grain to the other, or as we say, there is one to one correspondence of atoms at the boundary. As the atoms in the boundaries fit together well, coherent boundaries have low energies, and that is why all boundaries tend to become as coherent as possible.

Twin boundary is a coherent boundary of figure. Other examples of such boundaries are in widmanstatten structures, martensite, and in precipitates from supersaturated solid solutions. The important requirement for the formation of coherent boundary is that the atoms in the adjoining crystals must, at the common boundary (coherent), have closely similar patterns and spacings.

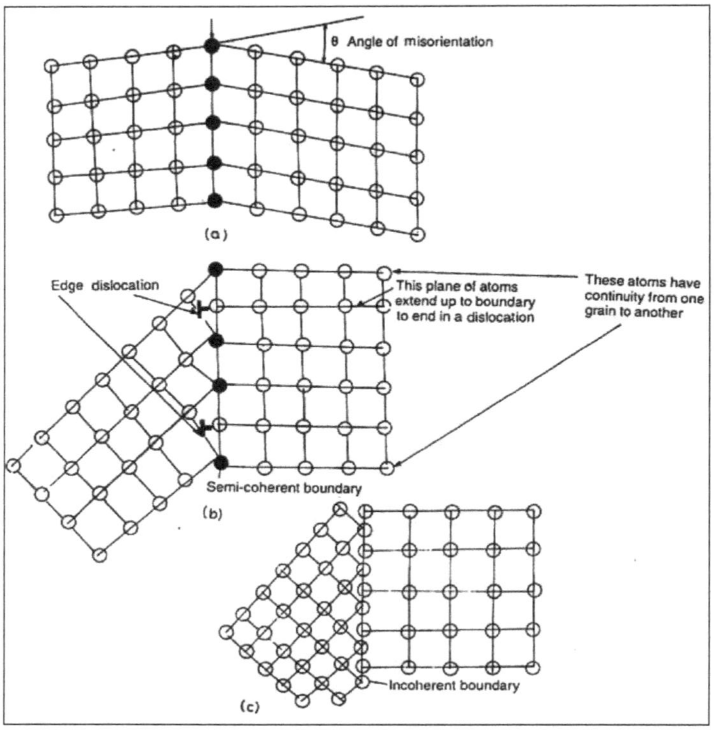

Schematic illustrations of coherent, semi-coherent and incoherent boundaries in a single phase poly-crystalline material. (a) Coherent boundary in a single phase poly-crystalline material. (b) Semi-coherent in a single phase polycrystalline material. (c) Incoherent boundary.

Small-angle boundary is a semi-coherent boundary. If the dislocations get introduced in a coherent boundary, it results in a semi-coherent boundary. There is continuity of atoms in some layers in both the adjacent grains, but ends up in a dislocation in other layers at the boundary.

By suitably and steadily increasing the density of dislocations in a boundary, (to represent increasing rotations of the crystals, or the boundary, or both) from orientation that produces coherent boundaries, a gradual change from coherent to semi-coherent to in-coherent boundaries can be obtained, figure illustrate this change. In the last extreme limit, the dislocations are too densely packed to overlap cores of neighbouring dislocations, to result in incoherent, or so called normal grain boundaries.

In multi-phase materials, when second phase forms in alloy crystals, then to get a coherent interface between the phases, the crystal structures of both phases have to be matched. If the size of precipitate is very small, it may adopt a crystal structure that allows it to be coherent with the matrix crystal, even if for matching the atomic bonds of two crystal structures at the boundary produce considerable elastic distortions in it, and its immediate surroundings.

The lattice constant of one phase increases and the other decreases. The thin disc shaped particles have least strain energy, and still be coherent. These thin tiny particles, which are coherent, can be thought of as solute-rich clusters in the parent matrix. In quenched and aged aluminium with 4 percent copper alloy, the tiny particles are copper rich discs about 100 A in diameter and 4 A thick on {100} planes of aluminium matrix with coherent interfaces.

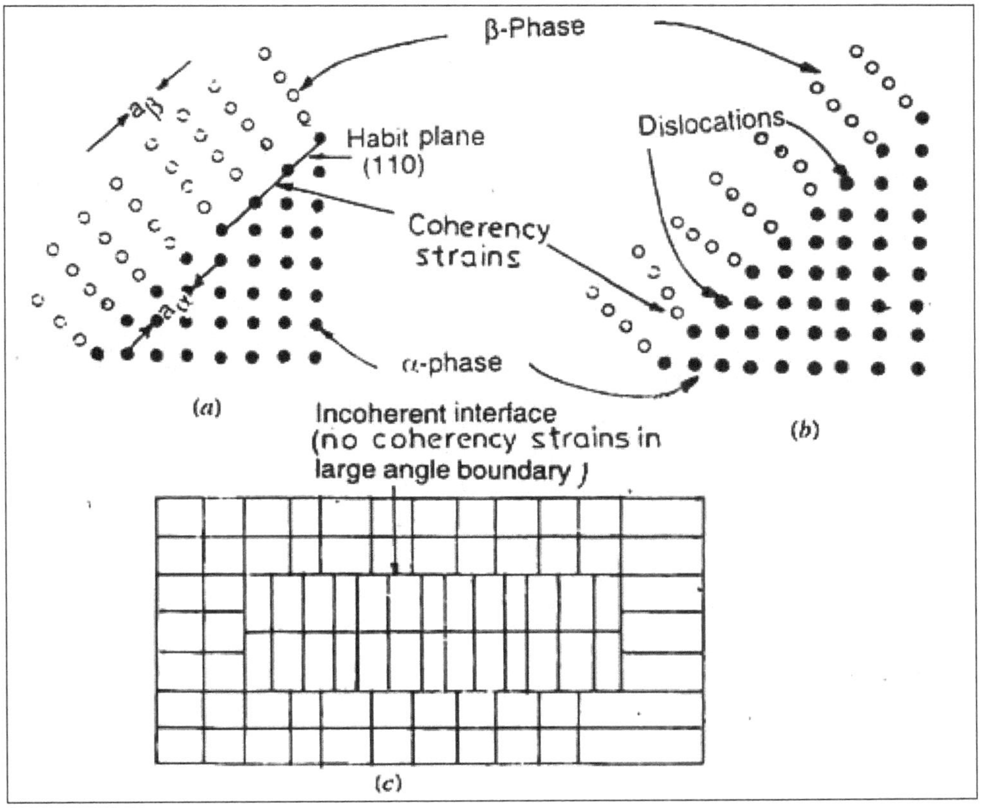

Figure shows coherent, semi-coherent and incoherent interfaces (boundaries) in poly-phase materials (Schematic). (a) A coherent interface with lattice strain due to the disregistry, $\delta = \frac{\alpha_\beta - \alpha_\alpha}{\alpha_\beta} < 0.1$.

(b) A semi-coherent interface with dislocations, which patially relieve the lattice strains due to disregistry, (c) Incoherent interface (no lattice strains in a large angle boundary.)

When such a coherent particle grows larger by obtaining atoms from matrix and crystal structure changes, it may break away (may be partly in the beginning) from its coherent attachments to the matrix in order to reduce the increasing elastic coherency strains, and adopts a more stable structure by introducing dislocations in its interface to make the interface semi-coherent, which on further change of the precipitate forms an incoherent interface.

Common Grain Boundaries

These boundaries are formed when two neighbouring grains, solidifying from liquid metal, or recrystallising, impinge on each other. The resulting boundary can be quite arbitrary because there are five degrees of freedom of such a boundary. The dislocation model of low-angle grain boundary illustrates that with the increase in angle of misorientation between the grains, the dislocations lie closer together.

At sufficiently high angles of misorientation, the individual dislocations lose their identity, because the cores of the dislocations overlap. The structure and the energy of such boundaries, called the large angle boundaries or common boundaries are difficult to be predicted. However, it is clear that grains on both sides of them are truly crystalline up to the boundary, which is a narrow transition region of imperfect order.

The grain boundaries are not necessarily smooth curvilinear surfaces, but with ledges and facets of macro as well as microscopic dimensions, particularly if the grains are held at-sufficiently high temperature to minimise the local elastic strains.

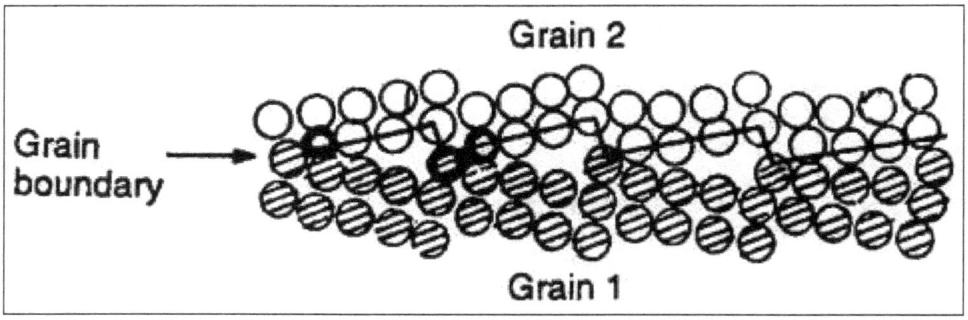

Ledges and facets in a grain boundary.

The energies of large-angle boundaries are usually estimated by thermal-etching. When a polycrystalline material is heated at high temperatures, thermal grooves form on the free surface, wherever a grain boundary intersects the surface perpendicularly.

This is an example of general effect of balancing the forces of surface tension at an interfacial triple junction, consisting of one grain boundary and free surfaces to the right and to the left of the point X.

If Υ_s is the free surface tension and Υ_b is surface energy of the boundary, then the two surface tensions of the free surface balance each other, but the vertically down component due to grain boundary remains unbalanced.

In order to balance this grain boundary energy, a groove must form as illustrated, figure (b) with a dihedral angle, θ between the two surface energies of free surfaces to satisfy the equation, figure (c).

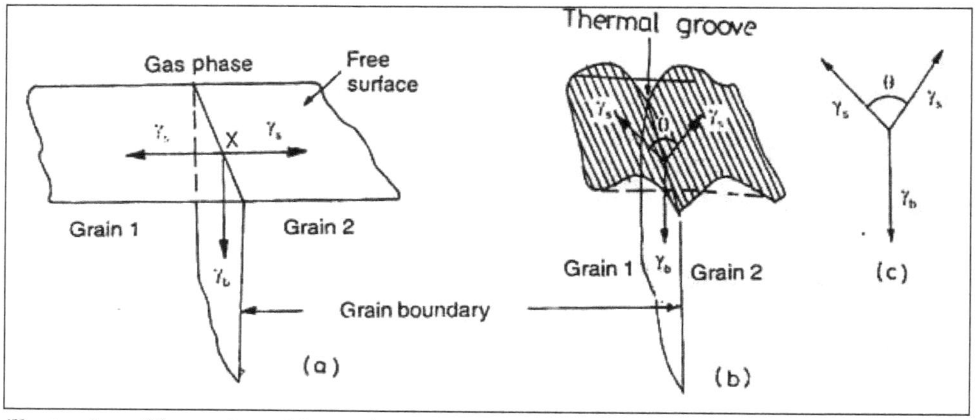

Schematic illustration of formation of thermal groove. (a) A grain boundary meets a free surface perpendicularly, (b) Thermal groove is a stable configuration at the junction of grain boundary with the surface. (c) Interfacial triple junction at groove base.

Diffusion and evaporation processes at high temperatures enable the necessary boundary movements to take place. It becomes possible to estimate Υ_b if θ and Υ_s are known. Actually, this equation gives free energy of a boundary, because the total energy is larger than this, due to extra contribution associated mainly with the vibrational entropy of atoms in the boundary. Table provides free energies of some large-angle boundaries. We find, $\Upsilon_b \approx 1/3\ \Upsilon_s$.

Table: Free energies of large-angle boundaries of some metals.

Υ_b, Jm^{-2}	Au	Cu	Fe	Ni	Sn
	0.36	.060	0.78	0.69	0.16

Grain boundaries have dislocations and vacancies in them. Infact, grain boundaries are source and sink of them. Soluble impurities strongly segregate to the grain boundaries to reduce their elastic misfit energies In fact, first few atomic percent of impurity as well as alloy atoms are invariably present at the grain boundaries as these can be accommodated better in imperfectly ordered grain boundaries.

In zone-refined metals, grain boundaries migrate readily at around 10-5m sec-1 at temperatures close to the melting point, and some movement has been also detected at 0.25 Tm.p. But soluble impurities at the grain boundaries decrease the movement of grain boundaries. The insoluble impurities have an even stronger effect in reducing grain boundary motion (also thus grain growth).

Problem: A thermal groove was observed on the surface of iron where the grain boundary intercepted it perpendicularly making a dihedral angle of 147°. The surface energy of iron is 1.49 Jm^{-2}. What is the grain boundary energy of iron?

Solution: We see in diagram -

$$\Upsilon_b = 2\ \Upsilon_s \cos \theta/2$$

$$= 2 \times 1.49 \times \cos 73.5^\circ$$

$$= 0.85\ Jm^{-2}$$

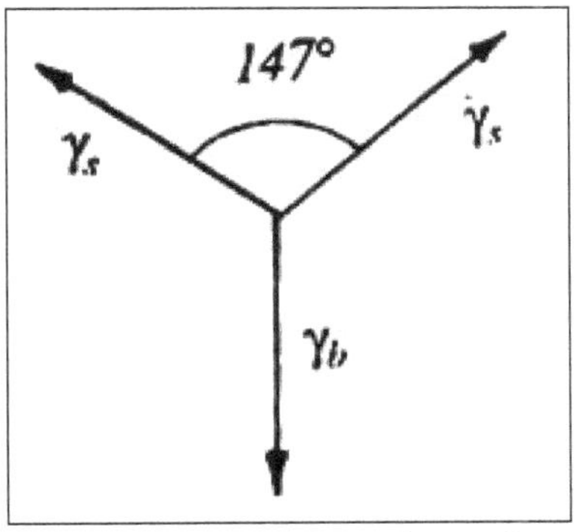

Interfacial triple junction at thermal groove.

Stacking-faults

Stacking-faults are planar surface defects produced by a fault (mistake) in the stacking sequence of atomic planes in crystals. The stacking sequence of planes of atoms on {111} planes in FCC crystal structure has been designated as ABCABC ..., whereas in HCP crystal structure on (0001) plane as ABABAB....

A fault in stacking sequence of say FCC crystal structure may be produced by plastic deformation, or by collapse of a large number of vacancies on one plane, or by agglomeration of a large number of interstitialies on a plane, etc.

Suppose a large number of vacancies collapse on a (111) plane, so that one plane of atoms is missing, say plane C is missing indicated by arrow:

$$\ldots ABC\ AB\overset{\downarrow}{C}\ ABC \ldots,$$

then the stacking sequence becomes:

... ABC AB ABC ...

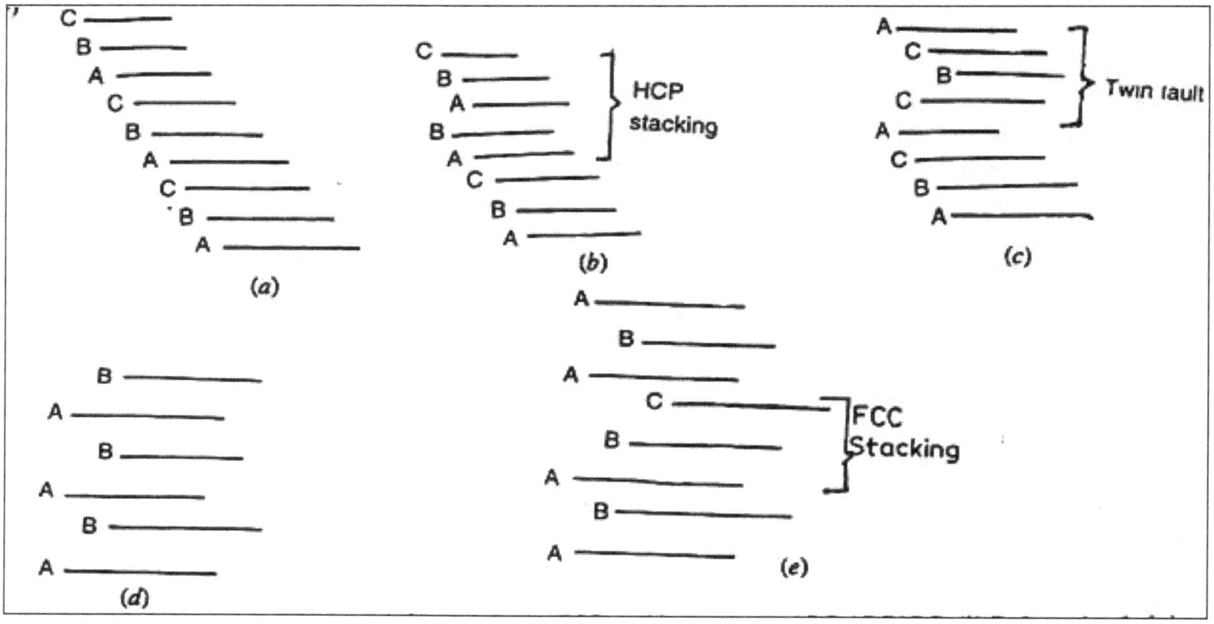

Stacking faults in FCC and HCP crystals. (a) FCC stacking sequence ABC ABCABC. (b) Deformation fault in FCC : AB-CAB : ABC. (c) Twin fault in FCC : ABCACBCABC (d) HCP stacking sequence. (e) FCC fault in HCP crystal.

There is now a fault in stacking, and this type of fault is called intrinsic stacking-fault as also seen in the formation of Frank loop by the internal collapse of vacancies. If we compare this sequence (underlined stacking sequence) with an HCP stacking sequence without any fault, that is, ... AB AB AB..., we find that stacking-fault contains four layers of an HCP stacking sequence.

Thus, formation of an intrinsic stacking-fault in an FCC metal is equivalent to the formation of a thin HCP region. This thin region is a surface defect called stacking-fault. The creation of a region with HCP stacking ABAB introduces a region with higher free energy than the FCC structure.

The coordination number in the faulted region is though twelve as in the unfaulted or perfect region, but the second-nearest-neighbour bonds in the faulted region are not of correct type for FCC perfect lattice. Thus, energy is associated for the creation of the stacking-fault, and is called stacking fault energy, which is normally of the order of 0.01 to 0.05 Jm^{-2}.

Table: Stacking Fault Width and Energies of Some Materials.

Metal	Stacking fault energy, γ_s, J/m^2	Equilibrium width of faulted region, d_s, A
Lead	0.025	-
Silver	0.025	18.7
Gold	0.050	5.7
Copper	0.080	6.8
Aluminium	0.200	1.5 (considered unextended)
Nickel	0.240	2.0
Brass	0.01	-
Stainless Steel 303	0.008	-

If some interstitialies agglomerate on a plane of atoms, addition of a plane of atom (at least in some area of it) takes place, It is because of this area of extra atoms as a part of plane that the stacking sequence is faulted, and such a stacking fault is called an extrinsic stacking fault. For example, HCP crystal having a stacking sequence of:

$$ABAB\overset{\downarrow}{A}B...$$

Is made to have an extra plane of atoms added, say, at the place of the arrow, then the atoms of this extra plane occupy a position different than positions of A or B, but say at C. At positions either A or B, the structure shall be highly distorted as then we shall have either position B on B or A on A. Now, at C positions, the stacking sequence becomes:

...ABABCAB ...

That is, an HCP crystal has a stacking-fault which contains three layers of FCC stacking sequence, and thus it shall require extra energy. Because of the additional lattice discontinuity associated with an extrinsic fault, it has higher energy than an intrinsic fault.

The introduction of an extrinsic fault in FCC crystal structure changes the stacking sequence from:

$$ABC\overset{\downarrow}{A}BCABC...$$

To

ABCACBCABC ...

By adding an extra plane of atoms at C positions in between A and B as indicated by arrow. The three layers ACB constitute a twin (of BCA). Thus, an extrinsic stacking-fault in FCC crystal is also called twin stacking fault.

This stacking fault in FCC metals can also be considered sub-microscopic twin of approximate atomic thickness. The stacking-fault energies of different FCC metals is responsible for their differential deformation behaviour. Metals in general with low-stacking-fault energies strain harden more rapidly, and twin easily on annealing.

Stacking-faults are easily detected after etching under an optical microscope, because of difference in the etching characteristics of faulted and parent structure.

Volume Defects

Volume defects in crystals are three-dimensional aggregates of atoms or vacancies. It is common to divide them into four classes in an imprecise classification that is based on a combination of the size and effect of the particle. The four categories are: precipitates, which are a fraction of a micron in size and decorate the crystal; second phase particles or dispersants, which vary in size from a fraction of a micron to the normal grain size (10-100μm), but are intentionally introduced into the microstructure; inclusions, which vary in size from a few microns to macroscopic dimensions, and are relatively large, undesirable particles that entered the system as dirt or formed by precipitation; and voids, which are holes in the solid formed by trapped gases or by the accumulation of vacancies.

Precipitates are small particles that are introduced into the matrix by solid state reactions. While precipitates are used for several purposes, their most common purpose is to increase the strength of structural alloys by acting as obstacles to the motion of dislocations. Their efficiency in doing this depends on their size, their internal properties, and their distribution through the lattice. However, their role in the microstructure is to modify the behavior of the matrix rather than to act as separate phases in their own right.

Dispersants are larger particles that behave as a second phase as well as influencing the behavior of the primary phase. They may be large precipitates, grains, or polygranular particles distributed through the microstructure. When a microstructure contains dispersants such properties as mechanical strength and electrical conductivity are some average of the properties of the dispersant phase and the parent.

Inclusions are foreign particles or large precipitate particles. They are usually undesirable constituents in the microstructure. For example, inclusions have a deleterious effect on the useful strength of structural alloys since they are preferential sites for failure. They are also often harmful in microelectronic devices since they disturb the geometry of the device by interfering in manufacturing, or alter its electrical properties by introducing undesirable properties of their own.

Voids (or pores) are caused by gases that are trapped during solidification or by vacancy condensation in the solid state. They are almost always undesirable defects. Their principal effect is to decrease mechanical strength and promote fracture at small loads.

References

- Alloys, science - 51812: britannica.com, Retrieved 19 May, 2019

- Schottky-defect, chemistry: byjus.com, Retrieved 14 August, 2019
- Frenkel-defect-point-defect-in-ionic-crystal, materials-science-defects: minaprem.com, Retrieved 25 January, 2019
- Difference-between-schottky-defect-and-frenkel-defect, materials-science: difference.minaprem.com, Retrieved 03 February, 2019
- Interstitial-defect-point-defect-defects-in-solid, materials-science: minaprem.com, Retrieved 05 March, 2019
- Interstitial-defect, materials-science: sciencedirect.com, Retrieved 18 August, 2019
- Interstitial-defect-point-defect-defects-in-solid, materials-science: minaprem.com, Retrieved 09 June, 2019
- Surface-defects-in-crystals-3-types-metallurgy, metallurgy-defects-metallurgy – 43279: engineeringenotes.com, Retrieved 25 August, 2019

Permissions

All chapters in this book are published with permission under the Creative Commons Attribution Share Alike License or equivalent. Every chapter published in this book has been scrutinized by our experts. Their significance has been extensively debated. The topics covered herein carry significant information for a comprehensive understanding. They may even be implemented as practical applications or may be referred to as a beginning point for further studies.

We would like to thank the editorial team for lending their expertise to make the book truly unique. They have played a crucial role in the development of this book. Without their invaluable contributions this book wouldn't have been possible. They have made vital efforts to compile up to date information on the varied aspects of this subject to make this book a valuable addition to the collection of many professionals and students.

This book was conceptualized with the vision of imparting up-to-date and integrated information in this field. To ensure the same, a matchless editorial board was set up. Every individual on the board went through rigorous rounds of assessment to prove their worth. After which they invested a large part of their time researching and compiling the most relevant data for our readers.

The editorial board has been involved in producing this book since its inception. They have spent rigorous hours researching and exploring the diverse topics which have resulted in the successful publishing of this book. They have passed on their knowledge of decades through this book. To expedite this challenging task, the publisher supported the team at every step. A small team of assistant editors was also appointed to further simplify the editing procedure and attain best results for the readers.

Apart from the editorial board, the designing team has also invested a significant amount of their time in understanding the subject and creating the most relevant covers. They scrutinized every image to scout for the most suitable representation of the subject and create an appropriate cover for the book.

The publishing team has been an ardent support to the editorial, designing and production team. Their endless efforts to recruit the best for this project, has resulted in the accomplishment of this book. They are a veteran in the field of academics and their pool of knowledge is as vast as their experience in printing. Their expertise and guidance has proved useful at every step. Their uncompromising quality standards have made this book an exceptional effort. Their encouragement from time to time has been an inspiration for everyone.

The publisher and the editorial board hope that this book will prove to be a valuable piece of knowledge for students, practitioners and scholars across the globe.

Index

www.ingramcontent.com/pod-product-compliance
Lightning Source LLC
LaVergne TN
LVHW080646210926
848485LV00003B/336